New Aspects of Quantum Electrodynamics

Akitomo Tachibana

New Aspects of Quantum Electrodynamics

Springer

Akitomo Tachibana
Department of Micro Engineering
Kyoto University
Kyoto, Japan

ISBN 978-981-10-3131-1 ISBN 978-981-10-3132-8 (eBook)
DOI 10.1007/978-981-10-3132-8

Library of Congress Control Number: 2016960671

Printed on acid-free paper

This Springer imprint is published by Springer Nature
The registered company is Springer Nature Singapore Pte Ltd.
The registered company address is: 152 Beach Road, #21-01/04 Gateway East, Singapore 189721, Singapore

Preface

Quantum Mechanics: 100 Years of Mystery Solved!

In the theoretical study of the application of quantum electrodynamics (QED), recent progress in research has led to solving the mystery (as Feynman said) involved in the foundation of quantum mechanics. Because this is a very big achievement, we will first note this breakthrough in the title of this preface, and later demonstrate the individual research outcomes.

QED is a relativistic quantum field theory, a quantum theory of photons with electrons, and is considered the most successful accurate theory we have, e.g., to explain the Lamb shift, the anomalous magnetic moment of the electrons, and so on using the Feynman diagram technique of the covariant perturbation approach. We will elaborate the non-perturbation approach in this book.

This book presents new aspects of QED from basic physics to physical chemistry with mathematical rigor. Topics covered include spin dynamics, chemical reactivity, the dual Cauchy problem, and more. Readers interested in modern applications of quantum field theory in nano-, bio-, and open systems will enjoy learning how the up-to-date quantum theory of radiation with matter works in the world of QED. In particular, chemical ideas restricted now to nonrelativistic quantum theory are shown to be unified and extended to relativistic quantum field theory that is basic to particle physics and cosmology: realization of the new-generation quantum theory. Readers are assumed to have a background equivalent to an undergraduate student's elementary knowledge in electromagnetism, quantum mechanics, chemistry, and mathematics. This book makes use of abundant figures to help the reader grasp ideas quickly, includes many equations to help the reader to follow the logic step-by-step, and provides an ample range of examples and references to facilitate in-depth learning.

I would like to thank Drs. Koichi Nakamura, Kentaro Doi, Masato Senami, Kazuhide Ichikawa, Ludwik Komorowski, Piotr Ordon, Andrzej Sokalski, Paweł Szarek, Irene Yarovsky, David Henry, Hansong Cheng, Akinori Fukushima, Yuji

Ikeda, Hiroo Nozaki, Masahiro Fukuda, and members of the Tachibana Laboratory in Kyoto University for their collaboration and producing some of the figures of numerical calculations.

Kyoto, Japan
31 August 2016

Akitomo Tachibana

Contents

Chapter 1
Basic Physics of QED

Abstract Basic physics of quantum electrodynamics (QED) is reviewed in comparison with quantum mechanics. Under external source of electromagnetic fields, charged particles can be accelerated by the Lorentz force. The Lorentz force is compensated by tension at any point of the Minkowski space-time. The tension is given by the divergence of internal self-stress tensor. The antisymmetric component of the stress tensor leads to spin torque and drives time evolution of electron spin. This is called the quantum electron spin vorticity principle. The spin torque can be compensated by a force called zeta force.

Keywords Alpha-oscillator theory • Chirality • Double slit • Dual Cauchy problem • Electromigration • Helicity • Measurement • Primary Rigged QED theory • Principle of equivalence • Response • Rigged QED theory • Spin torque • Spin vorticity • Spindle structure • Stress tensor • Tension • Zeta force • Zeta potential

1.1 Introduction

1.1.1 QED and Quantum Mechanics

In the Einstein special theory of relativity, a measurement of an "event" α is discussed on the Minkowski space-time. Let an event α be characterized in relativistic quantum field theory by a field operator $\widehat{F}(ct, x, y, z)$ at the Minkowski space-time coordinates (ct, x, y, z) as shown in Fig. 1.1. This is the standard framework of QED. In quantum mechanics, however, more operators $\widehat{x}, \widehat{y}$, and $\widehat{z}$ with $\widehat{F}(ct, \widehat{x}, \widehat{y}, \widehat{z})$ are required to discuss the measurement problem.

This additional expectation value problem of $\widehat{x}, \widehat{y}$, and $\widehat{z}$ with $\widehat{F}(ct, \widehat{x}, \widehat{y}, \widehat{z})$ in quantum mechanics may be viewed as "the icing on the cake" from that in QED. In QED, the Cartesian coordinates x, y, and z are merely the scale in inches or cm for $\widehat{F}(ct, x, y, z)$ and are not the objects of observation. In QED, the Cartesian coordinates x, y, and z are not observables nor canonical variables nor operators. So that in QED, we have no problem with the collapse of wave function nor classical

A. Tachibana, *New Aspects of Quantum Electrodynamics*,
DOI 10.1007/978-981-10-3132-8_1

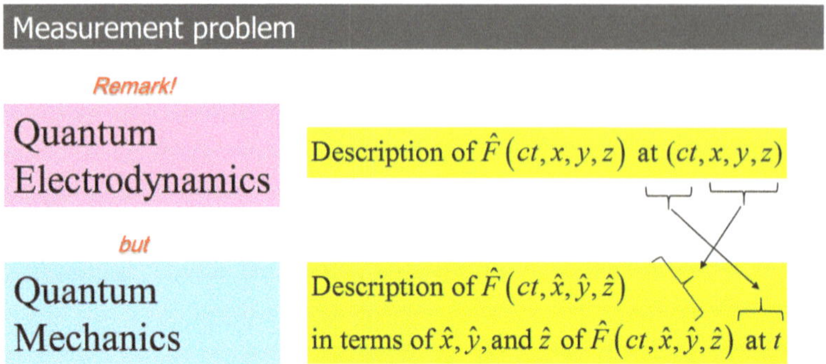

Fig. 1.1 Measurement of an "event" α in QED with the Minkowski space-time coordinates (ct, x, y, z) is different from that in quantum mechanics

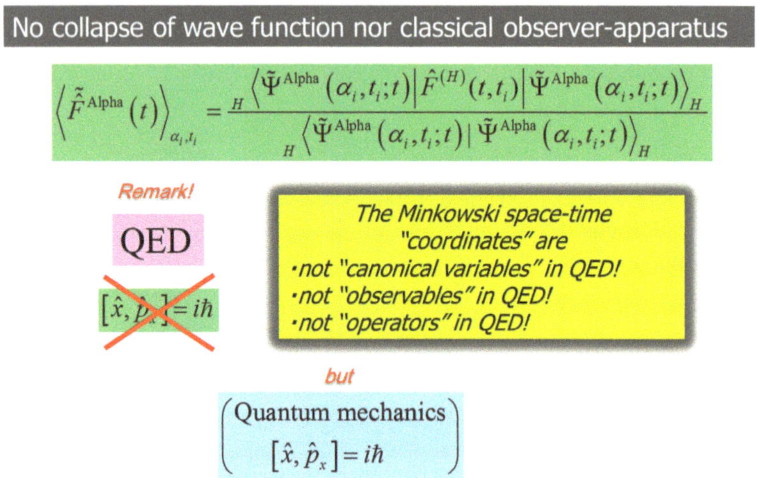

Fig. 1.2 Expectation value in QED is different from that in quantum mechanics. See Eq. (4.197) and Chap. 4 for further details with notation

observer–apparatus since the Cartesian coordinates x, y, and z are determined well before any discussion of measurement (see Fig. 1.2).

Historically, the foundation of quantum mechanics started with matrix mechanics by Heisenberg, Born, and Jordan and later with the physically equivalent wave mechanics by Schrödinger. In QED, the matrix mechanics is attributed to the field operator separated from wave mechanics. That's why we treat the uncertainty of measurement of the field operator $\widehat{F}(ct, x, y, z)$ in terms of the expectation value separated from wave mechanics in QED. QED is a relativistic quantum field theory and is considered the most successful accurate theory we have, e.g., to explain the Lamb shift, the anomalous magnetic moment of electron, and so on using the Feynman diagram technique of the covariant perturbation approach. We shall elaborate the non-perturbation approach in this book.

1.1.2 *The Most Beautiful Scientific Experiment*

Please refer to Fig. 1.3a. This is an experiment that has been done at Fundamental Research Laboratory, Hitachi Ltd. In 2002, in a vote by readers of Physics World magazine (is a member magazine of the IOP, the UK Institute of Physics), "the most beautiful scientific experiment" was coined to the selected quantum mechanics "double-slit experiment" (Crease 2006). Also shown in Fig. 1.3b is the double-slit experiment of photon that has been done at Hamamatsu Photonics, K.K.

Looking at the integrated data of the electron and photon spots, only discrete random spot as the number is low is observed. Gradually as the number increases double-slit phenomenon in which the quantum mechanics of the wave function is to be prophetic, probability distribution is emerging. But nobody has ever succeeded in proving that the quantum mechanics of wave function gives the precise distribution. As a matter of fact, nobody can (see Chap. 4). Not quantum mechanics but QED gives the correct answer (Tachibana 2016).

Double-slit experiments with elementary particles like electrons and photons have been carried out all over the world.

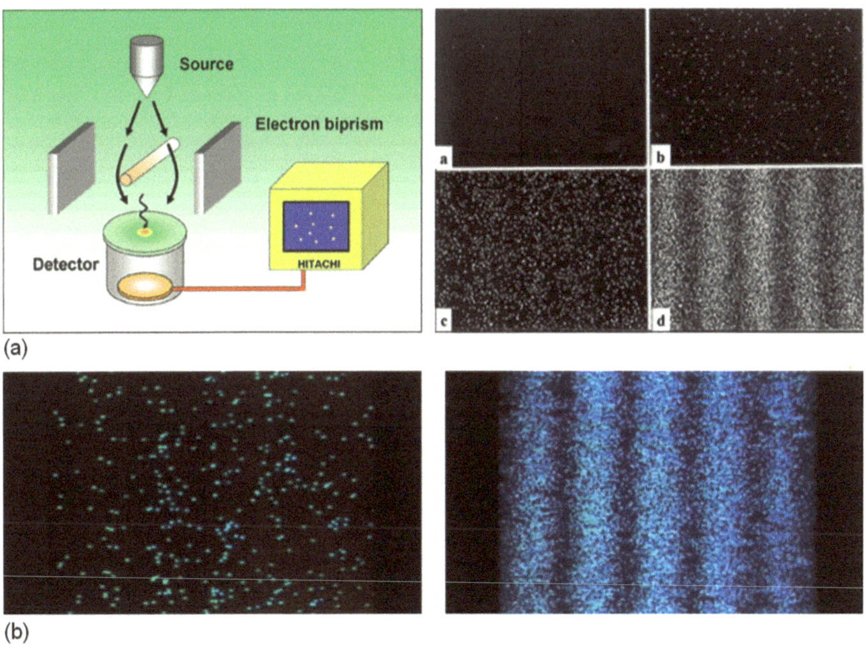

Fig. 1.3 Double-slit experiment of (**a**) electron (Reproduced from Hitachi, Ltd. http://www.hitachi.com/rd/portal/highlight/quantum/index.html) and (**b**) photon (Reproduced from Hamamatsu Photonics, K.K. http://photonterrace.net/en/photon/duality/)

1.1.3 Mystery of Quantum Mechanics

Quantum mechanics was founded around 100 years ago and is yet impossible to predict momentarily where each one electron or one photon should go. Copenhagen interpretation is accepted widely that Bohr, one of the founders of quantum mechanics, was to advocate. According to the Copenhagen interpretation, quantum mechanics of wave function is used in the description of the stochastic phenomenon. In contrast there is also a multi-world interpretation of Everett, one also an interpretation problem. The description of the phenomenon caused by the quantum mechanics continues to be a challenge that has also been left in the modern science.

As shown in Fig. 1.4, Einstein has pointed out the imperfections that lurking in the basic dynamics process of quantum mechanics, such as introduced in "God does not play dice" claimed that upon. Feynman, in his famous quantum mechanics textbook, described it as "the mystery of quantum mechanics" (Feynman et al. 1972).

1.1.4 New Theory

I recently found the "quantum mechanics of the mystery (Feynman says)" can be every moment predicted by QED (Tachibana 2016). As shown in Fig. 1.5, the key

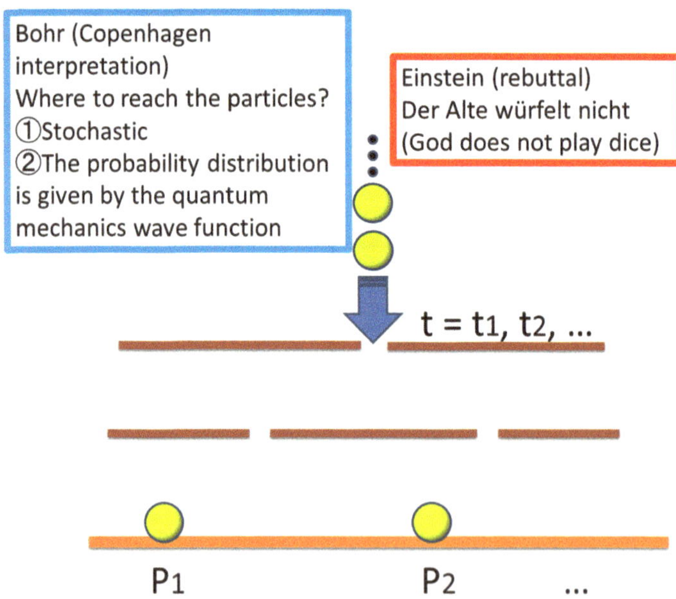

Fig. 1.4 Feynman said "the mystery of quantum mechanics" (Feynman et al. 1972)

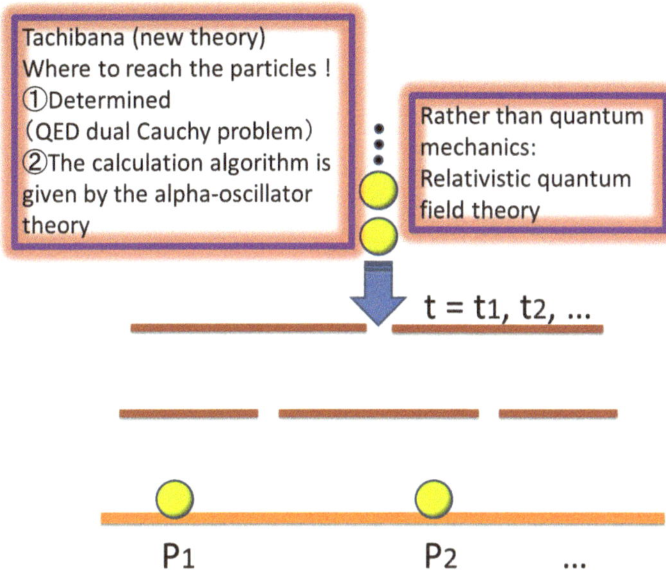

Fig. 1.5 Quantum mechanics 100 years of mystery is solved

lies in solving the dual Cauchy problem, the algorithm discovered was given by the alpha-oscillator theory.

Alpha-oscillator theory here was shown in the paper prior to this (Tachibana 2015). As has been presented in this series of papers, the new theory based on QED rather than quantum mechanics can predict a lot of interesting new phenomena including the double-slit phenomena (see Chap. 4 in details).

1.1.5 Survey of This Book

In Sect. 1.2 of this chapter, molecular dynamics of finite systems are unified with QED in terms of the Rigged QED theory. The Rigged QED theory is a non-perturbation approach to QED of finite systems. For finite systems, the local quantum physics of field theory has been reviewed (Haag 1992). We shall invoke the virial theorem (Landau and Lifshitz 1973) on the energetics of the finite systems in terms of the energy-momentum tensor in Sect. 2.2, Chap. 2.

Since QED is based on the theory of relativity, electron spin $\vec{s}_e$ should automatically be plugged in. What is new here is the quantum electron spin vorticity principle (see Fig. 1.6). What is vorticity of spin? It is defined by $\mathrm{rot}\vec{s}_e$ and it has the dimension of momentum. Interestingly, half the vorticity $\frac{1}{2}\mathrm{rot}\vec{s}_e(\vec{r})$ contributes to electron momentum. Why half?

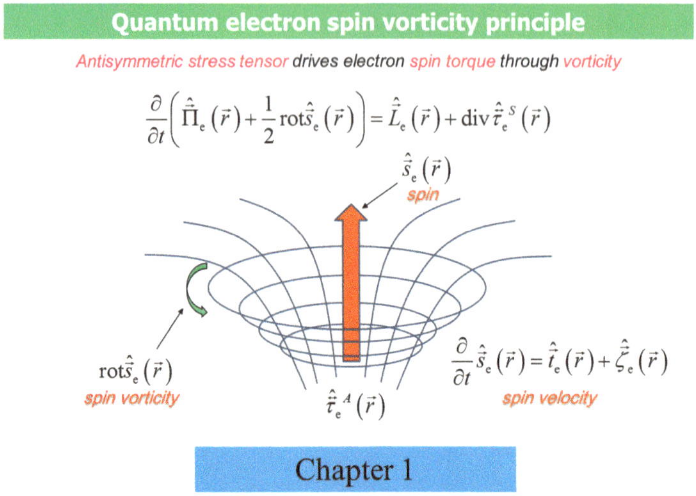

Fig. 1.6 Discovery of quantum electron spin vorticity principle

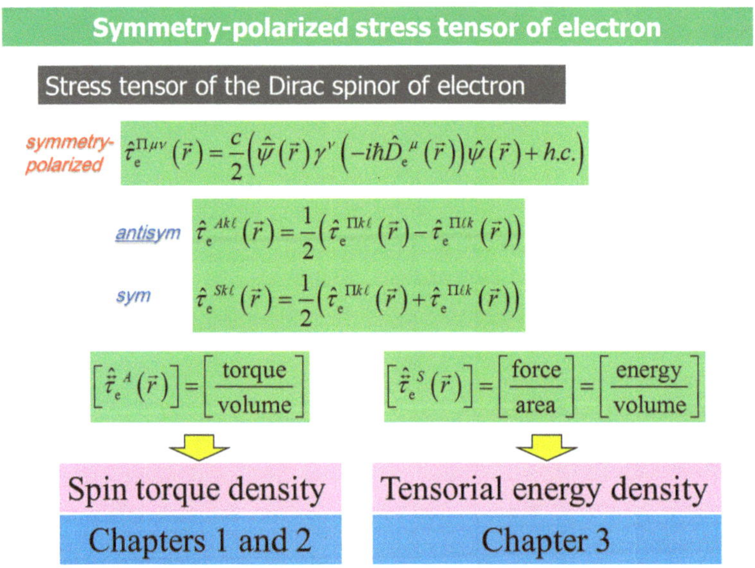

Fig. 1.7 Discovery of the symmetry-polarized stress tensor of electron

In Chap. 2, the reasoning "half" is found in the principle of equivalence (Tachibana 2012). The principle of equivalence requires that special relativity should apply in locally inertial frames and, in particular, that it should make no difference which locally inertial frame we choose at each point (Weinberg 1972). The mechanical framework of QED is represented as the symmetry of the stress tensor (see Fig. 1.7). So "energy-momentum tensor of QED" is the title of Chap. 2.

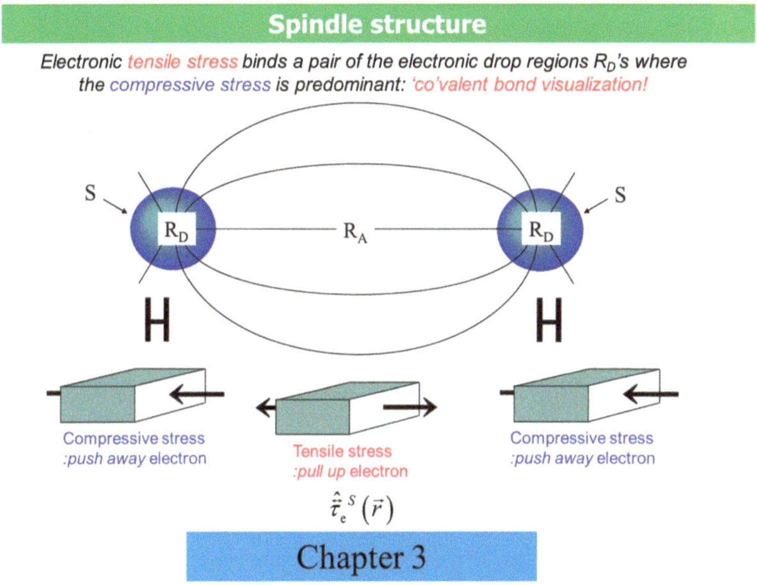

Fig. 1.8 Discovery of the spindle structure for the Lewis electron pairing as a tensile stress; novel local picture of covalency based on the electronic stress tensor

The energetics of the Rigged QED theory will also be discussed in terms of the energy-momentum tensor.

The symmetry of the stress tensor is polarized. The quantum electron spin vorticity principle is ascribed to the antisymmetric component. The antisymmetric component has the dimension of spin torque density. The symmetric component has the dimension of energy density. The tensorial energy density has prominent role as the spindle structure of covalency (Tachibana 2004).

"Chemical ideas of QED" is the title of Chap. 3 (see Fig. 1.8). Conventionally, the relativistic theory has been considered as only a slight correction for the interpretation in chemical phenomena. However, we have clarified that the Hamiltonian of QED, derived from the picture of "action through medium" based on the relativistic theory, gives a novel image of the chemical interaction even in the nonrelativistic limit (Tachibana 2013, 2014).

As a result, though the energy as an integrated value of the Hamiltonian of QED with respect to the whole space is equivalent to that of the usual ab initio Hamiltonian, conventional images of the chemical interaction based on "action at a distance" are replaced with the new images of them given by the picture of "action through medium" without exception.

In Chap. 4, quantum mechanics 100 years of mystery is solved. We shall apply the alpha-oscillator theory to QED, and find the dual Cauchy problem is the key to the solution (see Fig. 1.9). So "alpha-oscillator theory" is the title of Chap. 4.

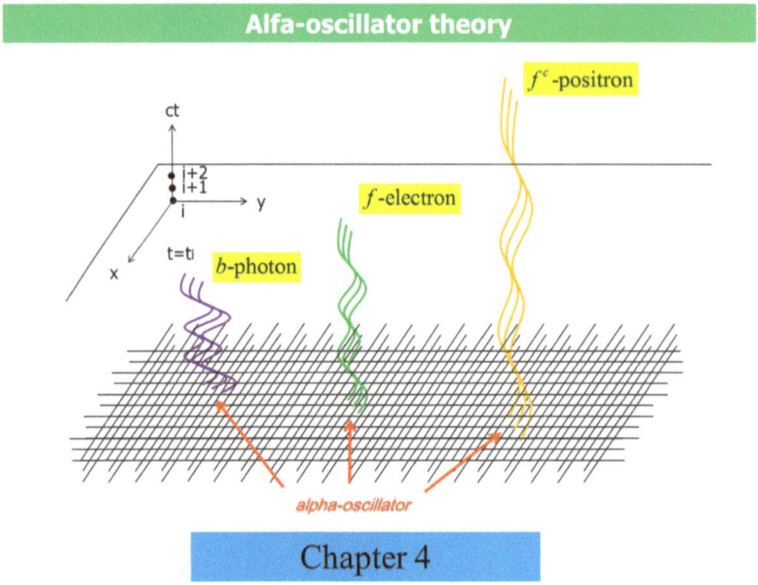

Fig. 1.9 Discovery of the alpha-oscillator theory

1.1.6 Quick Review of the Standard Theory of QED

Since this book intends to show only new aspects of QED, all the standard materials of QED are missing. The readers may consult the standard textbook for the conventional aspect of QED (Wigner 1939; Bargmann and Wigner 1948; Heitler 1954; Sakurai 1967; Bogoliubov et al. 1975; Itzykson and Zuber 1980; Berestetskii et al. 1982; Ryder 1985; Haag 1992; Nakanishi and Ojima 1990; Weinberg 1995; Peskin and Schroeder 1995; Greiner and Reinhardt 2009). Albeit duplicate, a quick review of the standard theory of QED will be introduced below.

In the standard model, the matter particles in general are spin-1/2 chiral fermions bound by gauge bosons satisfying the Poincare and gauge symmetries. The gauge fields of quantum chromodynamics (QCD) are reduced from the grand unified theory (GUT) as $SU(3)_c \times SU(2)_w \times U(1)_y \rightarrow SU(3)_c \times U(1)_{QED}$, where the Higgs field breaks the Weinberg–Salam electroweak gauge group $SU(2)_w \times U(1)_y$ down to $U(1)_{QED}$, but the color and charge symmetries remain intact. Quarks are bound by gluons $G_\mu{}^\ell$, while electron acquires its charge and mass through the Higgs mechanism with the Yukawa coupling, when massless photon A_μ as well as the massive bosons $Z_\mu{}^0$ and $W_\mu{}^\pm$ are emerging using the Glashow–Weinberg–Salam theory of spontaneously broken gauge symmetry. The gauge symmetry of the field theory is realized by the Becci–Rouet–Stora–Tyutin (BRST) symmetry of the Lagrangian $\partial_\theta \int d^4x \frac{1}{c}\widehat{L}(x) = 0$ where ∂_θ denotes the BRST operator. It follows that the physical content of the gauge theory is consistent with the cohomology of the BRST operator.

1.1.7 New-Generation Quantum Theory

QED allows the clamped-nuclei Hamiltonian, where the atomic nuclei are clamped in space and are treated as external static source of force for electrons. But in chemical reaction systems, the rearrangement of atomic configuration is of interest, and hence the dynamical treatments of atomic nuclei have been formulated by the Rigged QED theory.

Chemical ideas restricted now to nonrelativistic quantum theory may be unified and extended in the future to relativistic quantum field theory that is basic to particle physics and cosmology: realization of the new generation quantum theory. In order to accelerate this new trend, a topical symposium "New-Generation Quantum Theory—Particle Physics, Cosmology, and Chemistry" was organized aiming at mutually stimulating the cutting edge of basic theoretical approaches of quantum theory. Topics to be covered include, but not limited to, the cutting edge of basic theoretical approaches of quantum theory in particle physics, cosmology, and chemistry (see Fig. 1.10).

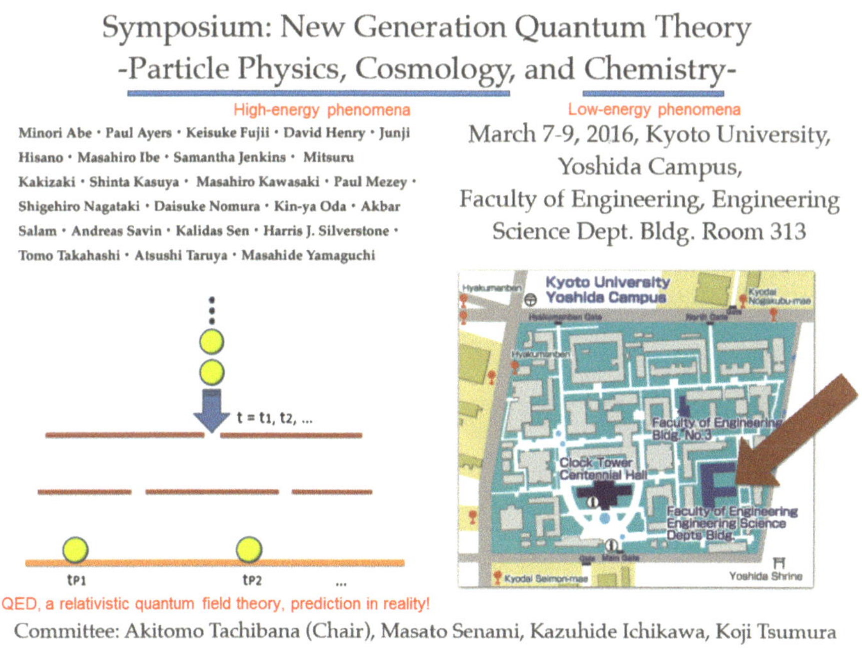

Fig. 1.10 Symposium: new-generation quantum theory—particle physics, cosmology, and chemistry (http://www.tachibana.kues.kyoto-u.ac.jp/symposium/01_top.html)

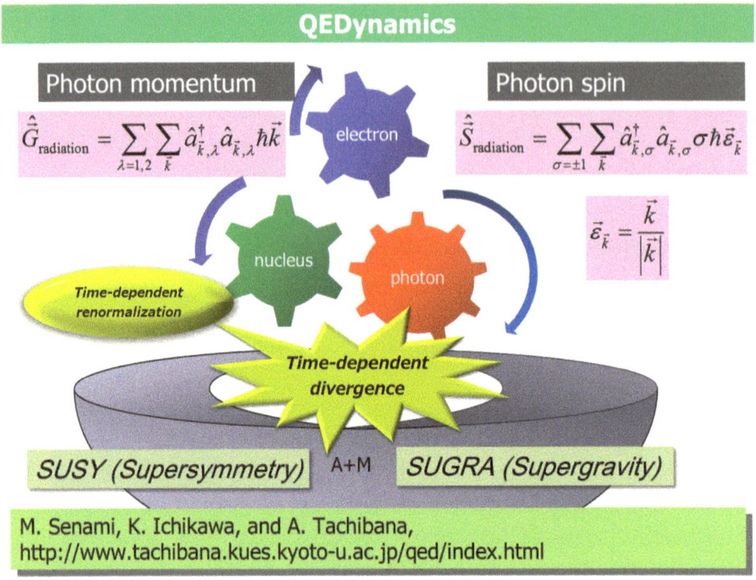

Fig. 1.11 QEDynamics: computer code for space-time-resolved non-perturbation simulation of the Rigged QED theory (http://www.tachibana.kues.kyoto-u.ac.jp/qed/index.html)

In this book, we have demonstrated preliminary numerical calculations of the Rigged QED theory. The numerical recipes with more advanced technics are all implemented in "QEDynamics," a computer code for space-time-resolved non-perturbation simulation of the Rigged QED theory (see Fig. 1.11).

We make every endeavor to realize the new generation quantum theory with the Rigged QED theory. The interested readers are encouraged to join us with the development of the new generation quantum theory.

1.1.8 Notation

The coordinate x with the contravariant components x^{μ} and the covariant components x_{μ} and the metric tensor $\eta_{\mu\nu} = \eta^{\mu\nu}$ of the Minkowski space-time, together with the inner product of two 4-vectors A and B written as $A{\cdot}B$ as well as the inner product of the Dirac gamma matrices γ^{μ} and a 4-vector A written as the Dirac slash $\not{A}$, are defined with the Euclidean inner product • as follows:

$$x^{\mu} = \left(x^0, x^k\right) = \left(x^0, x^1, x^2, x^3\right) = (ct, x, y, z) = (ct, \vec{r}) = (ct, \vec{x})$$

$$x_{\mu} = \eta_{\mu\nu}x^{\nu} = (x_0, x_k) = (x_0, x_1, x_2, x_3) = (ct, -x, -y, -z) = (ct, -\vec{r}) = (ct, -\vec{x})$$

$$\eta_{\mu\nu} = \begin{pmatrix} 1 & 0 & 0 & 0 \\ 0 & -1 & 0 & 0 \\ 0 & 0 & -1 & 0 \\ 0 & 0 & 0 & -1 \end{pmatrix} = \eta^{\mu\nu}, \quad \eta^{\mu\rho}\eta_{\rho\nu} = \delta^{\mu}{}_{\nu} = \begin{cases} 1, & \mu = \nu \\ 0, & \mu \neq \nu \end{cases}$$

$$A \cdot B = \eta_{\mu\nu}A^{\mu}B^{\nu} = A^0B^0 - \vec{A} \bullet \vec{B}, \quad \vec{A} \bullet \vec{B} = A_xB_x + A_yB_y + A_zB_z$$

$$\not{A} = \eta_{\mu\nu}\gamma^{\mu}A^{\nu} = \gamma^0 A^0 - \vec{\gamma} \bullet \vec{A}, \quad \vec{\gamma} \bullet \vec{A} = \gamma^1 A_x + \gamma^2 A_y + \gamma^3 A_z$$

where the Greek letter runs from 0 to 3 and the Latin from 1 to 3 and the Einstein summation convention is used. We use the chiral representation of γ^{μ} and the chiral matrix $\gamma_5 = -\gamma^5$ as

$$\gamma^{\mu} = (\gamma^0, \gamma^k) = (\gamma^0, \gamma^1, \gamma^2, \gamma^3), \quad \gamma_5 = i\gamma^0\gamma^1\gamma^2\gamma^3 = -\gamma^5$$

$$\{\gamma^{\mu}, \gamma^{\nu}\} = \gamma^{\mu}\gamma^{\nu} + \gamma^{\nu}\gamma^{\mu} = 2\eta^{\mu\nu}$$

$$\gamma^0 = \begin{pmatrix} 0 & 1 \\ 1 & 0 \end{pmatrix}, \quad \gamma^1 = \begin{pmatrix} 0 & -\sigma_x \\ \sigma_x & 0 \end{pmatrix}, \quad \gamma^2 = \begin{pmatrix} 0 & -\sigma_y \\ \sigma_y & 0 \end{pmatrix}, \quad \gamma^3 = \begin{pmatrix} 0 & -\sigma_z \\ \sigma_z & 0 \end{pmatrix}, \quad \gamma_5 = \begin{pmatrix} 1 & 0 \\ 0 & -1 \end{pmatrix}$$

with the Pauli matrices

$$\sigma_x = \begin{pmatrix} 0 & 1 \\ 1 & 0 \end{pmatrix}, \quad \sigma_y = \begin{pmatrix} 0 & -i \\ i & 0 \end{pmatrix}, \quad \sigma_z = \begin{pmatrix} 1 & 0 \\ 0 & -1 \end{pmatrix}$$

The Lorentz-invariant numerical tensor is the unit tensor $\delta^{\mu}{}_{\nu}$ and the Levi–Civita tensor (Landau and Lifshitz 1973)

$$\varepsilon^{\mu\nu\rho\sigma} = \begin{cases} 1, & \text{if } (\mu\nu\rho\sigma) \text{ is an even permutation of} (0123) \\ -1, & \text{if } (\mu\nu\rho\sigma) \text{ is an odd permutation of} (0123) \\ 0, & \text{otherwise} \end{cases}$$

$$\varepsilon^{0123} = 1, \quad \varepsilon_{0123} = -1$$

The gradient vectors are denoted as

$$\partial_{\mu} = \frac{\partial}{\partial x^{\mu}} = \left(\frac{\partial}{\partial x^0}, \frac{\partial}{\partial x^1}, \frac{\partial}{\partial x^2}, \frac{\partial}{\partial x^3}\right) = \left(\frac{1}{c}\frac{\partial}{\partial t}, \vec{\nabla}\right) = \left(\frac{1}{c}\frac{\partial}{\partial t}, \text{grad}\right)$$

$$\partial^{\mu} = \eta^{\mu\nu}\frac{\partial}{\partial x^{\nu}} = \left(\frac{\partial}{\partial x^0}, -\frac{\partial}{\partial x^1}, -\frac{\partial}{\partial x^2}, -\frac{\partial}{\partial x^3}\right) = \left(\frac{1}{c}\frac{\partial}{\partial t}, -\vec{\nabla}\right) = \left(\frac{1}{c}\frac{\partial}{\partial t}, -\text{grad}\right)$$

with the D'Alembertian

$$\Box = \partial^2 = \left(\frac{1}{c}\frac{\partial}{\partial t}\right)^2 - \Delta$$

and the Laplacian

$$\Delta = (\vec{\nabla})^2 = \frac{\partial^2}{\partial x^2} + \frac{\partial^2}{\partial y^2} + \frac{\partial^2}{\partial z^2}$$

The Dirac spinor $\psi(x)$ in the chiral representation is constructed by $\psi_R(x)$ with right-handed chirality and $\psi_L(x)$ with left-handed chirality as

$$\psi = \psi_{\text{chiral}} = \begin{pmatrix} \psi_R \\ \psi_L \end{pmatrix}, \quad \psi_R = \frac{1}{2}(1+\gamma_5)\psi, \quad \psi_L = \frac{1}{2}(1-\gamma_5)\psi$$

while in the Dirac representation, $\psi_A = (\psi_R + \psi_L)/\sqrt{2}$, $\psi_B = (\psi_R - \psi_L)/\sqrt{2}$. The spin density of electron is written in the bilinear covariant form as the axial vector (pseudovector):

$$\vec{s}(x) = \frac{1}{2}\hbar\overline{\psi}(x)\vec{\gamma}\gamma_5\psi(x) = \frac{1}{2}\hbar\vec{\sigma}(x)$$

$$\vec{\sigma}(x) = \vec{\sigma}_R(x) + \vec{\sigma}_L(x)$$

$$\vec{\sigma}_R(x) = {\psi_R}^\dagger(x)\vec{\sigma}\psi_R(x), \quad \vec{\sigma}_L(x) = {\psi_L}^\dagger(x)\vec{\sigma}\psi_L(x)$$

which is the spatial part of the third-rank antisymmetric tensor.

Also we have the chiral decomposition of electron current $j^\mu(x)$

$$\frac{1}{cq}j^\mu(x) = \overline{\psi}(x)\gamma^\mu\psi(x)$$

as

$$\frac{1}{cq}j^0(x) = \psi^\dagger(x)\psi(x) = N(x)$$

$$N(x) = N_R(x) + N_L(x)$$

$$N_R(x) = {\psi_R}^\dagger(x)\psi_R(x), \quad N_L(x) = {\psi_L}^\dagger(x)\psi_L(x)$$

and

$$\frac{1}{cq}\vec{j}(x) = \vec{\sigma}_R(x) - \vec{\sigma}_L(x)$$

Namely, the spatial part of the current density is given by the difference in the chiral parts of the spin density.

The chiral decomposition of the chiral current $j_5^{\mu}(x)$

$$\frac{1}{cq} j_5^{\mu}(x) = \overline{\psi}(x)\gamma^{\mu}\gamma_5\psi(x)$$

is found to be

$$\frac{1}{cq} j_5{}^0(x) = N_R(x) - N_L(x)$$

$$\frac{1}{cq} \vec{j}_5(x) = \vec{\sigma}(x) = \vec{\sigma}_R(x) + \vec{\sigma}_L(x)$$

Namely, the chiral charge density $j_5{}^0(x)$ is given by the difference in the chiral parts of the charge density, and the spatial part of the chiral current density $\vec{j}_5(x)$ is given by the spin density.

The spinor $\psi(x)$ in the chiral representation $\psi_{\text{chiral}}(x)$ is also constructed by the undotted spinor $\psi_R(x) = \xi^A(x)$ with right-handed chirality and the dotted spinor $\psi_L(x) = \eta_{\dot{U}}(x)$ with left-handed chirality as

$$\psi = \psi_{\text{chiral}} = \begin{pmatrix} \psi_R \\ \psi_L \end{pmatrix} = \begin{pmatrix} \xi^A \\ \eta_{\dot{U}} \end{pmatrix}$$

$$\xi^A = \begin{pmatrix} \xi^1 \\ \xi^2 \end{pmatrix}, \quad \eta_{\dot{U}} = \begin{pmatrix} \eta_{\dot{1}} \ \eta_{\dot{2}} \end{pmatrix}$$

The undotted and dotted capital Latin letters run from 1 to 2 and change position by using the antisymmetric matrix ε as

$$\xi_A = \xi^B \varepsilon_{BA}, \quad \eta^{\dot{U}} = \varepsilon^{\dot{U}\dot{V}} \eta_{\dot{V}}$$

$$\xi^A = \varepsilon^{AB} \xi_B, \quad \eta_{\dot{U}} = \eta^{\dot{V}} \varepsilon_{\dot{V}\dot{U}}$$

$$\varepsilon_{AB} = \begin{pmatrix} 0 & 1 \\ -1 & 0 \end{pmatrix} = \varepsilon^{AB}, \quad \varepsilon^{\dot{U}\dot{V}} = \begin{pmatrix} 0 & 1 \\ -1 & 0 \end{pmatrix} = \varepsilon_{\dot{U}\dot{V}}$$

where the Einstein summation convention is used.

The Pauli matrix σ with the contravariant components σ^{μ} and the covariant components σ_{μ} as

$$\sigma^{\mu} = \left(\sigma^0, \sigma^k\right) = \left(\sigma^0, \sigma^1, \sigma^2, \sigma^3\right) = \left(1, \sigma_x, \sigma_y, \sigma_z\right) = \left(1, \vec{\sigma}\right)$$

$$\sigma_{\mu} = \eta_{\mu\nu}\sigma^{\nu} = \left(\sigma_0, \sigma_k\right) = \left(\sigma_0, \sigma_1, \sigma_2, \sigma_3\right) = \left(1, -\sigma_x, -\sigma_y, -\sigma_z\right) = \left(1, -\vec{\sigma}\right)$$

(note the use of 1 as the unit matrix) are cast into the Misner–Thorne–Wheeler (MTW 1973) representation as

$$
\begin{aligned}
(\sigma_0)^{A\dot{U}} &= (\sigma^0)_{\dot{V}B} = \begin{pmatrix} 1 & 0 \\ 0 & 1 \end{pmatrix} = \sigma^0 \\
(\sigma_1)^{A\dot{U}} &= (\sigma^1)_{\dot{V}B} = \begin{pmatrix} 0 & 1 \\ 1 & 0 \end{pmatrix} = \sigma_x \\
(\sigma_2)^{A\dot{U}} &= (\sigma^2)_{\dot{V}B} = \begin{pmatrix} 0 & -i \\ i & 0 \end{pmatrix} = \sigma_y \\
(\sigma_3)^{A\dot{U}} &= (\sigma^3)_{\dot{V}B} = \begin{pmatrix} 1 & 0 \\ 0 & -1 \end{pmatrix} = \sigma_z
\end{aligned}
$$

Also, the Dirac gamma matrices γ^μ and the chiral matrix $\gamma_5 = -\gamma^5$ are given in the chiral representation using the MTW representation of the Pauli matrices as

$$
\begin{aligned}
\gamma^0 &= \begin{pmatrix} 0 & (\sigma_0)^{A\dot{U}} \\ (\sigma^0)_{\dot{V}B} & 0 \end{pmatrix} = \begin{pmatrix} 0 & \sigma^0 \\ \sigma^0 & 0 \end{pmatrix} = \begin{pmatrix} 0 & 1 \\ 1 & 0 \end{pmatrix} \\
\gamma^k &= \begin{pmatrix} 0 & -(\sigma_k)^{A\dot{U}} \\ (\sigma^k)_{\dot{V}B} & 0 \end{pmatrix} = \begin{pmatrix} 0 & -\sigma^k \\ \sigma^k & 0 \end{pmatrix} \\
\gamma_5 &= \begin{pmatrix} (\sigma^0)^A{}_B & 0 \\ 0 & -(\sigma^0)_{\dot{U}}{}^{\dot{V}} \end{pmatrix} = \begin{pmatrix} \sigma^0 & 0 \\ 0 & -\sigma^0 \end{pmatrix} = \begin{pmatrix} 1 & 0 \\ 0 & -1 \end{pmatrix} = -\gamma^5
\end{aligned}
$$

where the following MTW representation is found for the diagonal block

$$
\begin{aligned}
(\sigma^0)^A{}_B &= (\sigma^0)_{\dot{U}}{}^{\dot{V}} = \sigma^0 \\
(\sigma^1)^A{}_B &= (\sigma^1)_{\dot{U}}{}^{\dot{V}} = \sigma_x \\
(\sigma^2)^A{}_B &= (\sigma^2)_{\dot{U}}{}^{\dot{V}} = \sigma_y \\
(\sigma^3)^A{}_B &= (\sigma^3)_{\dot{U}}{}^{\dot{V}} = \sigma_z
\end{aligned}
$$

The Clifford algebra of the Dirac gamma matrices should be

$$
\{\gamma^\mu, \gamma^\nu\} = 2\eta^{\mu\nu} \begin{pmatrix} (\sigma^0)^A{}_B & 0 \\ 0 & (\sigma^0)_{\dot{U}}{}^{\dot{V}} \end{pmatrix} = 2\eta^{\mu\nu} \begin{pmatrix} 1 & 0 \\ 0 & 1 \end{pmatrix} = 2\eta^{\mu\nu}
$$

The charge conjugation matrix

$$
C = i\gamma^2\gamma^0 = \begin{pmatrix} \begin{pmatrix} 0 & -1 \\ 1 & 0 \end{pmatrix}^A{}_B & 0 \\ 0 & \begin{pmatrix} 0 & 1 \\ -1 & 0 \end{pmatrix}_{\dot{U}}{}^{\dot{V}} \end{pmatrix}
$$

with

$$C = C^* = -{}^tC = -C^{-1}, \quad C^\dagger C = 1, \quad CC = -1$$

transforms the Dirac gamma matrices as

$$C\gamma^\mu C - 1 = -{}^t\gamma^\mu$$
$$C\gamma_5 C^{-1} = {}^t\gamma 5 = \gamma_5$$

Complex conjugate ($c.c.$), transpose, the Hermitian conjugate ($h.c.$), and the Dirac conjugate matrix or operator A, are denoted as A^*, ${}^tA$, $A^\dagger = {}^tA^*$, and $\bar{A} = A^\dagger \gamma^0$ respectively.

We write

$$\{A, B\} = AB + BA = [A, B]_+; \quad [A, B] = AB - BA = [A, B]_-$$

Field variable F is denoted as

$$F(x) = F(ct, \vec{r}) = F(\vec{r}) = F(t)$$

where the dependence on the Minkowski space-time variables x^μ is frequently abbreviated if there arises no confusion.

The 3-vector external product is defined by using the Levi–Civita symbol as

$$\left(\vec{A} \times \vec{B}\right)^k = \varepsilon_{\ell nk} A^\ell B^n$$

$$\varepsilon_{\ell nk} = \begin{cases} 1, & \text{if } (\ell nk) \text{ is an even permutation of } (123) \\ -1, & \text{if } (\ell nk) \text{ is an odd permutation of } (123) \\ 0, & \text{otherwise} \end{cases}$$

$$\varepsilon_{123} = 1$$

For 3-vector $\vec{A}(x) = \begin{pmatrix} A_x(x) \\ A_y(x) \\ A_z(x) \end{pmatrix}$, the rot$\vec{A}(x)$ is defined by the rule

$$\text{rot}\vec{A}(x) = \vec{\nabla} \times \vec{A}(x) = \begin{pmatrix} \frac{\partial}{\partial y} A_z(x) - \frac{\partial}{\partial z} A_y(x) \\ \frac{\partial}{\partial z} A_x(x) - \frac{\partial}{\partial x} A_z(x) \\ \frac{\partial}{\partial x} A_y(x) - \frac{\partial}{\partial y} A_x(x) \end{pmatrix}$$

Likewise, for 3×3 tensor $\overleftrightarrow{T}(x) = \begin{pmatrix} T_{xx}(x) & T_{xy}(x) & T_{xz}(x) \\ T_{yx}(x) & T_{yy}(x) & T_{yz}(x) \\ T_{zx}(x) & T_{zy}(x) & T_{zz}(x) \end{pmatrix}$, the $\vec{r} \times \overleftrightarrow{T}(x)$ is defined by the rule

$$\vec{r} \times \overleftrightarrow{T}(x) = \begin{pmatrix} y\begin{pmatrix} T_{zx}(x) \\ T_{zy}(x) \\ T_{zz}(x) \end{pmatrix} - z\begin{pmatrix} T_{yx}(x) \\ T_{yy}(x) \\ T_{yz}(x) \end{pmatrix} \\ z\begin{pmatrix} T_{xx}(x) \\ T_{xy}(x) \\ T_{xz}(x) \end{pmatrix} - x\begin{pmatrix} T_{zx}(x) \\ T_{zy}(x) \\ T_{zz}(x) \end{pmatrix} \\ x\begin{pmatrix} T_{yx}(x) \\ T_{yy}(x) \\ T_{yz}(x) \end{pmatrix} - y\begin{pmatrix} T_{xx}(x) \\ T_{xy}(x) \\ T_{xz}(x) \end{pmatrix} \end{pmatrix}$$

The divergence of tensor density $T^{\cdot\cdot k\ell}(x)$ is defined by the rule

$$(\mathrm{div}T)^{\cdots k}(x) = \partial_\ell T^{\cdots k\ell}(x)$$

Whence

$$\vec{r} \times \mathrm{div}\overleftrightarrow{T}(x) = \begin{pmatrix} y\left(\mathrm{div}\overleftrightarrow{T}(x)\right)_z - z\left(\mathrm{div}\overleftrightarrow{T}(x)\right)_y \\ z\left(\mathrm{div}\overleftrightarrow{T}(x)\right)_x - x\left(\mathrm{div}\overleftrightarrow{T}(x)\right)_z \\ x\left(\mathrm{div}\overleftrightarrow{T}(x)\right)_y - y\left(\mathrm{div}\overleftrightarrow{T}(x)\right)_x \end{pmatrix} = \mathrm{div}\left(\vec{r} \times \overleftrightarrow{T}(x)\right)$$

The Kronecker delta symbol here is

$$\delta_{ij} = \begin{cases} 1, & i = j \\ 0, & \text{otherwise} \end{cases}$$

The generator $U(\Lambda, a)$ of the Poincaré group reduces to the infinitesimal transformation as

$$U(1+\omega,\varepsilon) = 1 - \frac{1}{2}i\omega_{\mu\nu}J^{\mu\nu}/\hbar + i\varepsilon_\mu P^\mu/\hbar + \cdots$$
$$J^{\mu\nu\dagger} = J^{\mu\nu} = -J^{\nu\mu}, \quad P^{\mu\dagger} = P^\mu$$
$$U(\Lambda,a)J^{\mu\nu}U^{-1}(\Lambda,a) = \Lambda_\rho{}^\mu\Lambda_\sigma{}^\nu(J^{\rho\sigma} - a^\rho P^\sigma + a^\sigma P^\rho)$$
$$U(\Lambda,a)P^\mu U^{-1}(\Lambda,a) = \Lambda_\rho{}^\mu P^\rho$$

leading to the Lie algebra

$$[P^\mu, P^\nu] = 0$$
$$[P^\mu, J^{\rho\sigma}] = i\hbar(\eta^{\mu\rho}P^\sigma - \eta^{\mu\sigma}P^\rho)$$
$$[J^{\mu\nu}, J^{\rho\sigma}] = i\hbar(\eta^{\nu\rho}J^{\mu\sigma} - \eta^{\mu\rho}J^{\nu\sigma} - \eta^{\mu\sigma}J^{\rho\nu} + \eta^{\nu\sigma}J^{\rho\mu})$$

The chiral spinor representation $D(\Lambda)$ of the Poincaré group reduces to the infinitesimal Lorentz transformation as

$$D(1+\omega) = 1 - \frac{1}{2} i\omega_{\mu\nu} J^{\mu\nu}/\hbar + \cdots$$

$$X = \begin{pmatrix} \xi^A \\ \eta_{\dot U} \end{pmatrix}, \quad X' = D(\Lambda)X = \begin{pmatrix} (\lambda_\xi \xi)^A \\ (\lambda_\eta \eta)_{\dot U} \end{pmatrix}, \quad D(\Lambda) = \begin{pmatrix} (\lambda_\xi)^A{}_B & 0 \\ 0 & (\lambda_\eta)_{\dot U}{}^{\dot V} \end{pmatrix}$$

$$(\lambda_\xi)^A{}_B (\sigma^\mu)^{B\dot V} \left(\lambda_\eta{}^{-1}\right)_{\dot V}{}^{\dot U} = (\lambda_\xi)^A{}_B (\sigma^\mu)^{B\dot V} \left(\lambda_\xi{}^\dagger\right)_{\dot V}{}^{\dot U} = \Lambda_\rho{}^\mu (\sigma^\rho)^{A\dot U}$$

$$(\lambda_\eta)_{\dot U}{}^{\dot V} (\sigma^\mu)_{\dot V B} \left(\lambda_\xi{}^{-1}\right)^B{}_A = (\lambda_\eta)_{\dot U}{}^{\dot V} (\sigma^\mu)_{\dot V B} \left(\lambda_\eta{}^\dagger\right)^B{}_A = \Lambda_\rho{}^\mu (\sigma^\rho)_{\dot U A}$$

$$\begin{aligned} D(\Lambda)\gamma^\mu D^{-1}(\Lambda) &= \Lambda_\rho{}^\mu \gamma^\rho \\ D(\Lambda) J^{\mu\nu} D^{-1}(\Lambda) &= \Lambda_\rho{}^\mu \Lambda_\sigma{}^\nu J^{\rho\sigma} \end{aligned}$$

leading to

$$J^{\mu\nu} = \frac{1}{4} i\hbar [\gamma^\mu, \gamma^\nu] = \frac{1}{2}\hbar\sigma^{\mu\nu}, \quad \sigma^{\mu\nu} = \frac{1}{2} i [\gamma^\mu, \gamma^\nu]$$

$$J^{k\ell} = \frac{1}{2}\hbar\varepsilon_{k\ell m} \begin{pmatrix} (\sigma^m)^A{}_B & 0 \\ 0 & (\sigma^m)_{\dot U}{}^{\dot V} \end{pmatrix} = \frac{1}{2}\hbar\varepsilon_{k\ell m} \begin{pmatrix} \sigma^m & 0 \\ 0 & \sigma^m \end{pmatrix}$$

$$\vec J = \left(J^{23}, J^{31}, J^{12}\right) = \frac{1}{2}\hbar \begin{pmatrix} \vec\sigma & 0 \\ 0 & \vec\sigma \end{pmatrix}$$

$$J^{k0} = \frac{1}{2} i\hbar \begin{pmatrix} -(\sigma^k)^A{}_B & 0 \\ 0 & (\sigma^k)_{\dot U}{}^{\dot V} \end{pmatrix} = \frac{1}{2} i\hbar \begin{pmatrix} -\sigma^k & 0 \\ 0 & \sigma^k \end{pmatrix}$$

$$\vec K = \left(J^{10}, J^{20}, J^{30}\right) = \frac{1}{2} i\hbar \begin{pmatrix} -\vec\sigma & 0 \\ 0 & \vec\sigma \end{pmatrix}$$

So that

$$\left[J^k, J^\ell\right] = i\hbar\varepsilon_{k\ell n} J^n$$

$$\left(J^1 \pm iJ^2\right)_{m'm} = \delta_{m'm} \sqrt{(j \mp m)(j \pm m + 1)}\hbar$$

$$\left(J^3\right)_{m'm} = \delta_{m'm} m\hbar$$

$$-\left(\vec J\right)^*_{m',m} = (-)^{m-m'} \left(\vec J\right)_{-m,-m'}$$

$$\left[J^k, \frac{H}{c}\right] = 0, \quad \left[J^k, P^\ell\right] = i\hbar\varepsilon_{k\ell n} P^n$$

$$\left[K^k, K^\ell\right] = -i\hbar\varepsilon_{k\ell n} J^n$$

$$\left[K^k, J^\ell\right] = \left[J^k, K^\ell\right] = i\hbar\varepsilon_{k\ell n} K^n$$

$$\left[K^k, \frac{H}{c}\right] = i\hbar P^k, \quad \left[K^k, P^\ell\right] = i\hbar\delta_{k\ell} \frac{H}{c}$$

and

$$\vec{A} = \frac{1}{2}(\vec{J} + i\vec{K}), \quad \vec{B} = \frac{1}{2}(\vec{J} - i\vec{K})$$
$$\vec{J} = \vec{A} + \vec{B}, \quad \vec{K} = -i\left(\vec{A} - \vec{B}\right)$$
$$\left[A^k, A^\ell\right] = i\hbar\varepsilon_{k\ell n}A^n$$
$$\left[B^k, B^\ell\right] = i\hbar\varepsilon_{k\ell n}B^n$$
$$\left[\vec{A}, \vec{B}\right] = 0$$

The Levi–Civita connection is defined as

$$\left\{ {\lambda \atop \mu\nu} \right\} = \frac{1}{2}g^{\lambda\rho}\left(\partial_\mu g_{\nu\rho} + \partial_\nu g_{\mu\rho} - \partial_\rho g_{\mu\nu}\right)$$

More general connection $\Gamma^\lambda_{\mu\nu}$ to define the covariant derivative ";" of the Lorentz vector A_μ

$$A_{\mu;\nu} = \partial_\nu A_\mu - \Gamma^\lambda_{\mu\nu}A_\lambda$$

is used to define the curvature

$$A_{\mu;\nu;\sigma} - A_{\mu;\sigma;\nu} = A_\rho R^\rho{}_{\mu\nu\sigma}$$

with the Riemann-Christoffel curvature tensor $R^\rho{}_{\mu\nu\sigma}$

$$R^\rho{}_{\mu\nu\sigma} = \partial_\nu\Gamma^\rho_{\mu\sigma} - \partial_\sigma\Gamma^\rho_{\mu\nu} + \Gamma^\rho_{\lambda\nu}\Gamma^\lambda_{\mu\sigma} - \Gamma^\rho_{\lambda\sigma}\Gamma^\lambda_{\mu\nu}$$

The Ricci tensor defined as

$$R_{\mu\nu} = R^\rho{}_{\mu\nu\rho}$$

is used to define the scalar curvature

$$R = R^\mu{}_\mu$$

The Gaussian unit of electromagnetism is used with the elementary charge e and the speed c of light in vacuum. When appropriate for numerical demonstration, atomic unit is used unless otherwise stated explicitly. The suffixes "e" for electron and "^" for operator are suppressed if there arises no confusion.

1.2 Rigged QED Theory

1.2.1 Underlying History of the Rigged QED Theory

It seems to have been already known to Kepler that sunlight should have given some kind of mechanical action on the tail of comet. Since a technique for observing the diffraction of sunlight has not yet been developed in his time, it seems it was mainstream of the time to consider sunlight from a mechanical point of view (see Fig. 1.12).

Afterward, by Newton who decomposed the sunlight through a prism, the sunlight was cast a particle theory of light, and the wave theory of light was not of reasonable shape (see Fig. 1.13).

Young was a stir, as he proved experimentally the double-slit phenomenon of sunlight from the analogy of a wave of water (see Fig. 1.14).

Since then, the wave theory of sunlight became dominant over the particle theory. On the other hand, Maxwell unified the electrostatic force of Coulomb and the electromagnetic induction of Faraday (see Fig. 1.15) and discovered the wave equation of sunlight with electricity and magnetism.

Fig. 1.12 The comet Hale–Bopp, seen here over the Joshua Tree National Park in Southern California on the evening of 28 March 1997, has both a *blue* ion tail and a *white* dust tail. Whereas the ion tail is carried away by the "solar wind" of charged particles from the Sun's atmosphere, the dust tail is pushed by the radiation pressure of the sunlight. The momentum transfer in this second case is weaker than that in the first, resulting in the splitting of the tails. This view of sunlight pressure dates back to the age of Kepler (Reproduced from Leonhardt 2006)

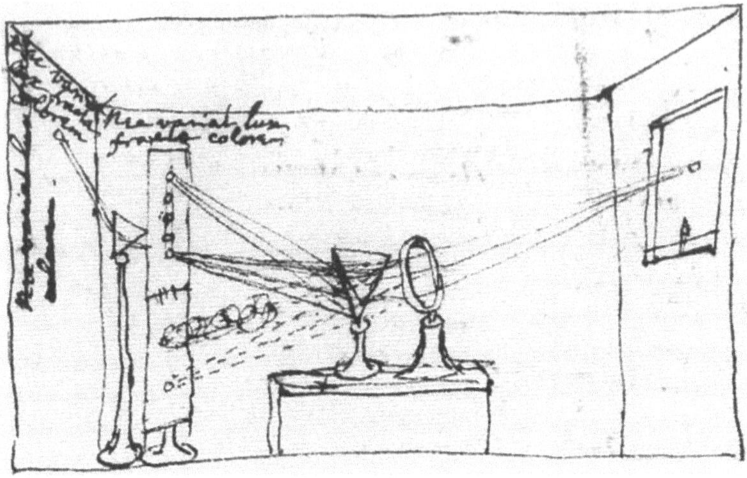

Fig. 1.13 Newton's sketch of his crucial experiment (1672), demonstrating corpuscular theory of sunlight (Reproduced from Fara 2015)

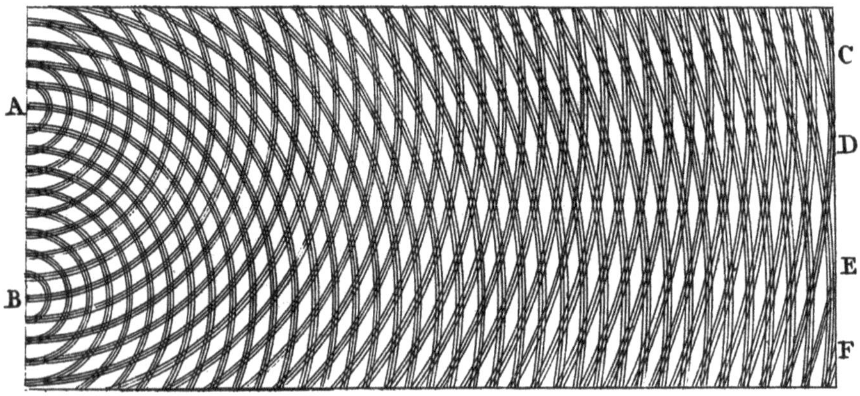

Fig. 1.14 Young's sketch (1807) of double-slit interference of sunlight in analogy of water waves (Reproduced from Rothman 2003)

The theory of electromagnetism by Maxwell is invariant under the Lorentz transformation and has been formulated as the special theory of relativity by Einstein (Jackson 1998).

The quantum theory of electromagnetism is QED (Weinberg 1995). According to the theory of electromagnetism, moving charged particle modifies the associated electromagnetic field with speed of light. The varying electromagnetic field propagates through space and gives impetus to other moving charged particle according to the Lorentz force. This demonstrates "action through medium" as the field theoretical nature of the electromagnetic interaction of charged particles. QED provides quantum mechanical framework for the field theoretical "action through

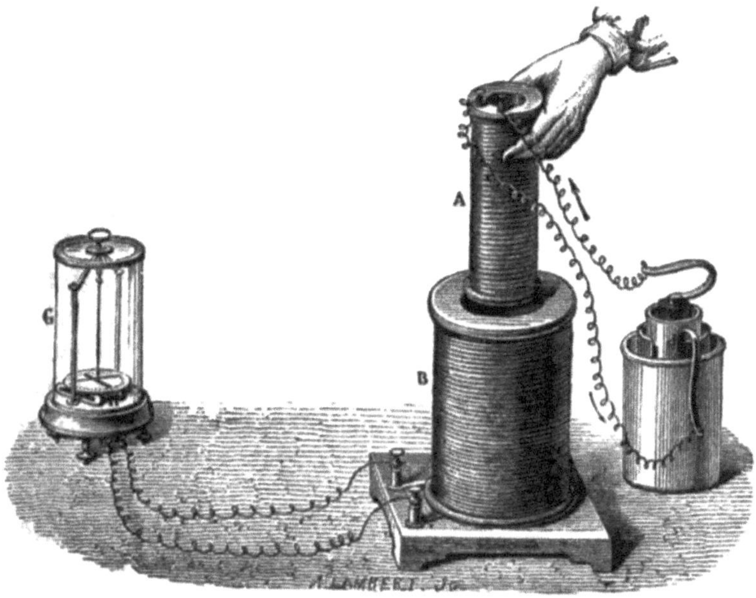

Fig. 1.15 Faraday's experiment (1831) showing electromagnetic induction (Reproduced from Poyser 1892)

medium" as the fundamental law of electrons interacting with atomic nuclei and radiation field of photons. QED gives rise to tension of stress tensor over and above the Lorentz force (Tachibana 2003). Poincare devised to attach stress tensor the equilibrium state of the charged particles that interact with the electromagnetic field (Jackson 1998).

For light atoms the speed of electrons is much slower than the speed of light, and hence we rely upon quantum mechanical electrostatic Coulomb law described by conventional nonrelativistic ab initio electrostatic Hamiltonian (Tachibana 2001). The Coulomb law demonstrates "action at a distance" as the electrostatic nature of the conventional interaction of charged particles. In contrast to the correct "action through medium" in QED, the conventional electrostatic interaction is instantaneous since the speed of light is eventually infinite for "action at a distance" in the Coulomb law. The interaction of charged particles with radiation field is then treated under the assumption that the Lorentz invariance is lost in the conventional treatment of the ab initio Hamiltonian.

Conventional QED theory assumes clamped-nuclei Hamiltonian in finite systems (Weinberg 1995), where the atomic nuclei are clamped in space and are treated as external static source of force for electrons. But in chemical reaction systems, the rearrangement of atomic configuration is of primary interest, and hence the dynamical treatments of atomic nuclei often play an important role.

We shall hence elaborate the incorporation of the kinetic energy density of atomic nuclei into the general framework of QED and present the general theory of the field energy density in finite chemical reaction systems. The atomic nucleus is assumed to be treated as a Schrödinger field. This is called the Rigged QED theory.

1.2.2 Basic Physics of the Rigged QED Theory

In the Rigged QED theory, atomic nucleus is plugged in as a Schrödinger field which is not a classical observer–apparatus. This is an essential advantage of the Rigged QED theory. In the conventional QED (Weinberg 1995), atomic nucleus is often implicitly treated as external potential, which is nothing but classical observer–apparatus. In quantum mechanics, moreover, the presence of classical observer–apparatus is mandatory. In the Rigged QED theory, we are free from the measurement problem of the Minkowski space-time coordinates, since we do not invoke the concept of the classical observer–apparatus. Even though we use the Schrödinger field for the nuclear dynamics, the space-time coordinates are merely the scale in inches or cm, not operators. In Chap. 2, the Schrödinger field for electron is derived in the primary Rigged QED theory. Again in the primary Rigged QED theory, the space-time coordinates are merely the scale in inches or cm, not the operators. In the primary Rigged QED theory also, we are free from the measurement problem of the Minkowski space-time coordinates, since we do not invoke the concept of the classical observer–apparatus.

We use the virial theorem (Landau and Lifshitz 1975) for the finite-system energetics of the Rigged QED theory and the primary Rigged QED theory as formulated so in Sect. 2.2, Chap. 2. For the sake of simplicity, the primary Rigged QED theory is used for all the numerical calculations of wave functions in this book unless otherwise stated explicitly. The wave functions are twofold: one for the expansion functions of fields and another for the ket vectors (see Chap. 4 in details). Atomic symbol is used to illustrate that the center of the wave function is localized around there. We use preliminary wave packets of electrons and nuclei centered around the atomic symbol. Albeit preliminary, it is based on the underlying physics shown in Sect. 1.1.1. Namely, it does not mean that we are working on the Born–Oppenheimer adiabatic approximation of quantum mechanics.

The equations of motion of fields are obtained using standard variation principle. The variation principle is made to be invariant under the Abelian $U(1)_{\mathrm{QED}}$ gauge transformation. This is the gauge principle of the Rigged QED theory. The stress tensor of the Rigged QED theory appears in the equation of motion of fields.

The Rigged QED theory is gauge invariant and preserves translational and rotational symmetry but violates the Poincare symmetry. This is because the presence of the Schrödinger fields violates the Lorentz invariance of the Lagrangian density. If we neglect the Schrödinger fields, then we recover the conventional QED with the Poincare symmetry as well as the gauge symmetry. The canonical quantization is performed with the gauge-invariant Lagrangian density operator using the Coulomb gauge at the starting point (see Fig. 1.16)

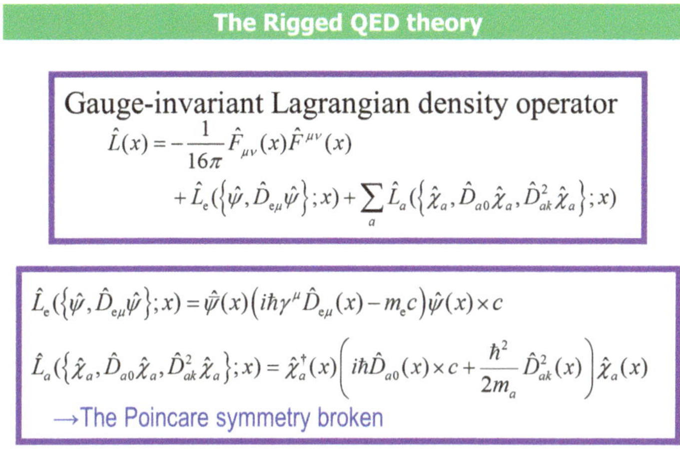

Fig. 1.16 The gauge-invariant Rigged QED theory Lagrangian density operator

$$\begin{aligned}\widehat{L}(x) &= -\frac{1}{16\pi}\widehat{F}_{\mu\nu}(x)\widehat{F}^{\mu\nu}(x) \\ &+\widehat{L}_e\left(\left\{\widehat{\psi}, \widehat{D}_{e\mu}\widehat{\psi}\right\}; x\right) + \sum_a \widehat{L}_a\left(\left\{\widehat{\chi}_a, \widehat{D}_{a0}\widehat{\chi}_a, \widehat{\vec{D}}^2_a\widehat{\chi}_a\right\}; x\right)\end{aligned} \tag{1.1}$$

where $\widehat{L}_e\left(\left\{\widehat{\psi}, \widehat{D}_{e\mu}\widehat{\psi}\right\}; x\right)$ is the Lagrangian density operator of electron

$$\widehat{L}_e\left(\left\{\widehat{\psi}, \widehat{D}_{e\mu}\widehat{\psi}\right\}; x\right) = c\widehat{\overline{\psi}}(x)\left(i\hbar\gamma^\mu\widehat{D}_{e\mu}(x) - m_e c\right)\widehat{\psi}(x) \tag{1.2}$$

and $\widehat{L}_a\left(\left\{\widehat{\chi}_a, \widehat{D}_{a0}\widehat{\chi}_a, \widehat{\vec{D}}^2_a\widehat{\chi}_a\right\}; x\right)$ is the Lagrangian density operator of a'th atomic nucleus:

$$\widehat{L}_a\left(\left\{\widehat{\chi}_a, \widehat{D}_{a0}\widehat{\chi}_a, \widehat{\vec{D}}^2_a\widehat{\chi}_a\right\}; x\right) = \widehat{\chi}^\dagger_a(x)\left(i\hbar\widehat{D}_{a0}(x) \times \mathrm{c} + \frac{\hbar^2}{2m_a}\widehat{\vec{D}}^2_a(x)\right)\widehat{\chi}_a(x) \tag{1.3}$$

The $\widehat{D}_{\alpha\mu}(x)$, where $\alpha = \mathrm{e}$ stands for electron and $\alpha = a$ stands for a'th atomic nucleus, is the covariant derivative operator using the Abelian gauge potential operator $\widehat{A}_\mu(x)$

$$\widehat{D}_{\alpha\mu}(x) = \partial_\mu + i\frac{q_\alpha}{\hbar c}\widehat{A}_\mu(x), \quad q_\alpha = Z_\alpha e \tag{1.4}$$

with m_e and $Z_e = -1$ being the mass and charge number of electron and m_a and Z_a being the mass and charge number of the a'th atomic nucleus, respectively. The nuclear spin is plugged in ad hoc. The canonical quantization rule of the Schrödinger field is anti-commutation relationship for fermions and commutation relationship for bosons.

The theory is invariant under gauge transformation

$$\widehat{\psi}(x) \to \widehat{\psi}^{(\theta)}(x) = \exp\left(i\frac{q_e}{\hbar c}\widehat{\theta}(x)\right)\widehat{\psi}(x) \tag{1.5}$$

$$\widehat{\chi}_a(x) \to \widehat{\chi}_a{}^{(\theta)}(x) = \exp\left(i\frac{q_a}{\hbar c}\widehat{\theta}(x)\right)\widehat{\chi}_a(x) \tag{1.6}$$

$$\widehat{A}_\mu(x) \to \widehat{A}_\mu{}^{(\theta)}(x) = \widehat{A}_\mu(x) - \partial_\mu\widehat{\theta}(x) \tag{1.7}$$

The gauge potential operators stand for

$$\widehat{A}^\mu(x) = \left(\widehat{\phi}(x), \widehat{\vec{A}}(x)\right) \tag{1.8}$$

The covariant derivative operators satisfy

$$\left[\widehat{D}_{\alpha\mu}(x), \widehat{D}_{\alpha\nu}(x)\right] = i\frac{q_\alpha}{\hbar c}\widehat{F}_{\mu\nu}(x) \tag{1.9}$$

with the gauge field operators $\widehat{F}_{\mu\nu}(x)$ defined as

$$\begin{aligned}\widehat{F}_{\mu\nu}(x) &= \partial_\mu\widehat{A}_\nu(x) - \partial_\nu\widehat{A}_\mu(x)\\ &= \begin{pmatrix} 0 & \widehat{E}_x(x) & \widehat{E}_y(x) & \widehat{E}_z(x) \\ -\widehat{E}_x(x) & 0 & -\widehat{B}_z(x) & \widehat{B}_y(x) \\ -\widehat{E}_y(x) & \widehat{B}_z(x) & 0 & -\widehat{B}_x(x) \\ -\widehat{E}_z(x) & -\widehat{B}_y(x) & \widehat{B}_x(x) & 0 \end{pmatrix}\end{aligned} \tag{1.10}$$

Then, the electric field operator $\widehat{\vec{E}}(x)$ and the magnetic field operator $\widehat{\vec{B}}(x)$ are given by using the Coulomb gauge potentials $\widehat{A}_\mu(x)$ as

$$\widehat{\vec{E}}(x) = -\mathrm{grad}\widehat{A}_0(x) - \frac{1}{c}\frac{\partial\widehat{\vec{A}}(x)}{\partial t}, \quad \widehat{\vec{B}}(x) = \mathrm{rot}\widehat{\vec{A}}(x), \quad \mathrm{div}\widehat{\vec{A}}(x) = 0 \tag{1.11}$$

1.2.3 The Maxwell Equations

The Rigged QED theory Maxwell equations of motion are found for the electromagnetic fields

$$\mathrm{rot}\widehat{\vec{E}}(x) + \frac{1}{c}\frac{\partial\widehat{\vec{B}}(x)}{\partial t} = 0 \tag{1.12}$$

The Rigged QED theory

The Rigged Maxwell equations of motion

$$\mathrm{rot}\hat{\vec{E}}(x)+\frac{1}{c}\frac{\partial\hat{\vec{B}}(x)}{\partial t}=0,\qquad \mathrm{div}\hat{\vec{B}}(x)=0$$

$$\mathrm{div}\hat{\vec{E}}(x)=4\pi\hat{\rho}(x),\qquad \mathrm{rot}\hat{\vec{B}}(x)-\frac{1}{c}\frac{\partial\hat{\vec{E}}(x)}{\partial t}=\frac{4\pi}{c}\hat{\vec{j}}(x)$$

Electric field and magnetic flux density

$$\hat{\vec{E}}(x)=-\mathrm{grad}\hat{\varphi}(x)-\frac{1}{c}\frac{\partial\hat{\vec{A}}(x)}{\partial t},\quad \hat{\vec{B}}(x)=\mathrm{rot}\hat{\vec{A}}(x),\quad \mathrm{div}\hat{\vec{A}}(x)=0$$

Fig. 1.17 The Rigged Maxwell equations with the electromagnetic fields

$$\mathrm{div}\widehat{\vec{B}}(x)=0 \tag{1.13}$$

$$\mathrm{div}\widehat{\vec{E}}(x)=4\pi\widehat{\rho}(x) \tag{1.14}$$

$$\mathrm{rot}\widehat{\vec{B}}(x)-\frac{1}{c}\frac{\partial\widehat{\vec{E}}(x)}{\partial t}=\frac{4\pi}{c}\widehat{\vec{j}}(x) \tag{1.15}$$

where $\widehat{\rho}(x)$ is the charge density operator and $\widehat{\vec{j}}(x)$ is the current density operator (see Fig. 1.17).

1.2.4 The Dirac–Schrödinger Equations

The Rigged QED theory Dirac equation of motion for the Dirac spinor field is

$$i\hbar\gamma^{\mu}\widehat{D}_{e\mu}(x)\widehat{\psi}(x)=m_{e}c\widehat{\psi}(x) \tag{1.16}$$

$$-i\hbar\left(\widehat{D}_{e\mu}(x)\widehat{\psi}(x)\right)^{\dagger}\gamma^{0}\gamma^{\mu}=m_{e}c\widehat{\overline{\psi}}(x) \tag{1.17}$$

Likewise, the Schrödinger equation of motion for the Schrödinger field is (see Fig. 1.18)

$$i\hbar\frac{\partial}{\partial t}\widehat{\chi}_{a}(x)=-\frac{\hbar^{2}}{2m_{a}}\widehat{\vec{D}}_{a}^{2}(x)\widehat{\chi}_{a}(x)+q_{a}\widehat{A}_{0}(x)\widehat{\chi}_{a}(x) \tag{1.18}$$

$$-i\hbar\frac{\partial}{\partial t}\widehat{\chi}_{a}^{\dagger}(x)=-\frac{\hbar^{2}}{2m_{a}}\left(\widehat{\vec{D}}_{a}^{2}(x)\widehat{\chi}_{a}(x)\right)^{\dagger}+q_{a}\left(\widehat{A}_{0}(x)\widehat{\chi}_{a}(x)\right)^{\dagger} \tag{1.19}$$

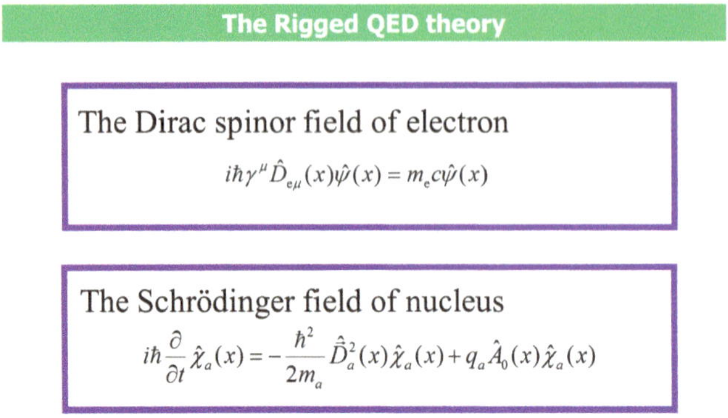

Fig. 1.18 The Rigged equations of motion with the Dirac and Schrödinger fields

1.2.5 Continuity Equations

It is easy to find that charge and current satisfy the Rigged QED theory continuity equation

$$\partial_{\mu}\hat{j}^{\mu}(x) = 0, \ \hat{j}^{\mu}(x) = \left(c\hat{\rho}(x), \hat{\vec{j}}(x)\right) \tag{1.20}$$

$$\frac{\partial}{\partial t}\hat{\rho}(x) + \mathrm{div}\hat{\vec{j}}(x) = 0 \tag{1.21}$$

where $\hat{\rho}(x)$ is the charge density operator and $\hat{\vec{j}}(x)$ is the charge current density operator (see Fig. 1.19).

The components satisfy

$$\partial_{\mu}\hat{j}_{\alpha}{}^{\mu}(x) = 0, \ \hat{j}_{\alpha}{}^{\mu}(x) = \left(c\hat{\rho}_{\alpha}(x), \hat{\vec{j}}_{\alpha}(x)\right) \tag{1.22}$$

$$\frac{\partial}{\partial t}\hat{\rho}_{\mathrm{e}}(x) + \mathrm{div}\hat{\vec{j}}_{\mathrm{e}}(x) = 0 \tag{1.23}$$

$$\frac{\partial}{\partial t}\hat{\rho}_{a}(x) + \mathrm{div}\hat{\vec{j}}_{a}(x) = 0 \tag{1.24}$$

The $\hat{\rho}(x)$ is decomposed into

$$\hat{\rho}(x) = \hat{\rho}_{\mathrm{e}}(x) + \sum_{a}\hat{\rho}_{a}(x) = \sum_{\alpha}\hat{\rho}_{\alpha}(x) \tag{1.25}$$

$$\hat{\rho}_{\alpha}(x) = q_{\alpha}\hat{N}_{\alpha}(x) \tag{1.26}$$

The Rigged QED theory

The Rigged charge and current density operators

$$\hat{\rho}(x) = \hat{\rho}_{\rm e}(x) + \sum_a \hat{\rho}_a(x), \quad \hat{\rho}_a(x) = q_a \hat{N}_a(x)$$

$$\hat{\vec{j}}(x) = \hat{\vec{j}}_{\rm e}(x) + \sum_a \hat{\vec{j}}_a(x), \quad \hat{\vec{j}}_a(x) = q_a \hat{\vec{S}}_a(x)$$

Nuclear density operator

$$\hat{N}_a(x) = \hat{\chi}_a^\dagger(x)\hat{\chi}_a(x)$$

Nuclear probability current density operator

$$\hat{S}_a^k(x) = \frac{1}{2m_a}\left(-i\hbar\hat{\chi}_a^\dagger(x)\hat{D}_{ak}(x)\hat{\chi}_a(x) + h.c.\right)$$

Fig. 1.19 Charge and current densities for the Rigged continuity equations

where $\hat{\rho}_{\rm e}(x)$ is the electronic charge density operator and $\hat{\rho}_a(x)$ is the charge density operator of a'th atomic nucleus and where $\hat{N}_{\rm e}(x)$ and $\hat{N}_a(x)$ are the position probability density operator of electron and a'th atomic nucleus, respectively:

$$\hat{N}_{\rm e}(x) = \hat{\bar{\psi}}(x)\gamma^0\hat{\psi}(x) \tag{1.27}$$

$$\hat{N}_a(x) = \hat{\chi}_a^\dagger(x)\hat{\chi}_a(x) \tag{1.28}$$

The $\hat{\vec{j}}(x)$ is decomposed into

$$\hat{\vec{j}}(x) = \hat{\vec{j}}_{\rm e}(x) + \sum_a \hat{\vec{j}}_a(x) = \sum_\alpha \hat{\vec{j}}_\alpha(x) \tag{1.29}$$

$$\hat{\vec{j}}_\alpha(x) = q_\alpha \hat{\vec{v}}_\alpha(x) \tag{1.30}$$

where $\hat{\vec{j}}_{\rm e}(x)$ is the electronic charge current density operator and $\hat{\vec{j}}_a(x)$ is the charge current density operator of a'th atomic nucleus and $\hat{\vec{v}}_\alpha(x)$ denotes the velocity density operator:

$$\hat{\vec{v}}_{\rm e}(x) = c\hat{\bar{\psi}}(x)\vec{\gamma}\hat{\psi}(x) \tag{1.31}$$

$$\hat{\vec{v}}_a(x) = \frac{1}{2m_a}\left(i\hbar\hat{\chi}_a^\dagger(x)\hat{\vec{D}}_a(x)\hat{\chi}_a(x) + h.c.\right) \tag{1.32}$$

By the Gordon decomposition, we have

$$\widehat{\vec{v}}_e(x) = \frac{1}{2m_e}\left(i\hbar\widehat{\overline{\psi}}(x)\widehat{\vec{D}}_e(x)\widehat{\psi}(x) - i\hbar\left(\widehat{\vec{D}}_e(x)\widehat{\psi}(x)\right)^{\dagger}\gamma^0\cdot\widehat{\psi}(x)\right)$$
$$+\frac{\hbar}{2m_e}\mathrm{rot}\left(\widehat{\overline{\psi}}(x)\vec{\sigma}\widehat{\psi}(x)\right) - \frac{i\hbar}{2m_e}\frac{\partial}{\partial t}\left(\widehat{\overline{\psi}}(x)\gamma^0\vec{\gamma}\widehat{\psi}(x)\right) \tag{1.33}$$

The $\widehat{\vec{v}}_a(x)$ may also be written as the flux density operator $\widehat{\vec{S}}_a(x)$ as follows:

$$\widehat{\vec{v}}_a(x) = \widehat{\vec{S}}_a(x) \tag{1.34}$$

1.2.6 The Lorentz Force and Stress Tensors

Under external source of electromagnetic fields, charged particles can be accelerated by the Lorentz force. In the Rigged QED theory, the tension density given by the divergence of stress tensor density $\widehat{\tau}_e^{\Pi}\mu\nu(x)$ acts as the counter force to the Lorentz force. Pauli in quantum mechanical context formulated the differential force law derived from the divergence relations applied to the energy-momentum tensor under general situations in the presence of electromagnetic fields (Pauli 1933), while the basic idea dates back to Schrödinger (1927). Moreover, the antisymmetric part $\widehat{\tau}_e^{A}\mu\nu(x)$ has unique physical meaning of spin torque density (Tachibana 2010). The spin torque density can be compensated by a force density, called zeta force density (Tachibana 2012). The symmetric part $\widehat{\tau}_e^{S}\mu\nu(x)$ has the physical meaning of tensorial energy density (see Fig. 1.20).

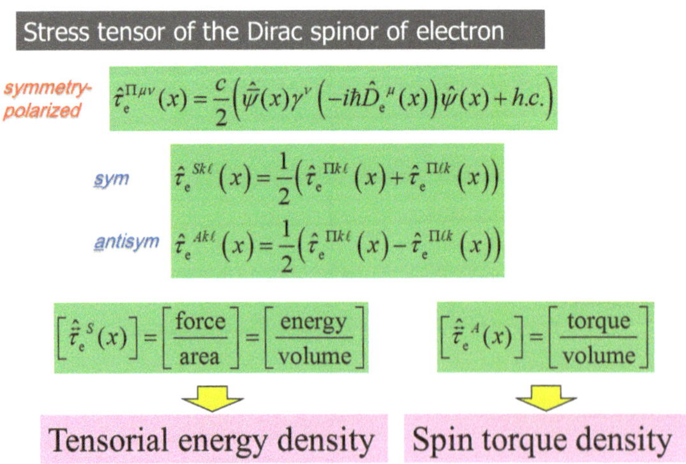

Fig. 1.20 Symmetry of the stress tensor of electron

The origin of the sunlight pressure (see Fig. 1.12) is represented by the Poynting electromagnetic field momentum density operator

$$\widehat{\vec{G}}(x) = \frac{1}{4\pi c}\widehat{\vec{E}}(x) \times \widehat{\vec{B}}(x) \tag{1.35}$$

It satisfies the equation of motion

$$\frac{\partial}{\partial t}\frac{1}{2}\left(\widehat{\vec{G}}(x) + \widehat{\vec{G}}^{\dagger}(x)\right) = -\frac{1}{2}\left(\widehat{\vec{L}}(x) + \widehat{\vec{L}}^{\dagger}(x)\right) - \mathrm{div}\widehat{\overleftrightarrow{\sigma}}(x) \tag{1.36}$$

where $\widehat{\overleftrightarrow{\sigma}}(x)$ is the Maxwell stress tensor density operator

$$\begin{aligned}
\widehat{\sigma}^{ij}(x) &= \frac{1}{8\pi}\left(\widehat{\vec{E}}^2(x)\delta_{ij} - \left(\widehat{E}^i(x)\widehat{E}^j(x) + \widehat{E}^j(x)\widehat{E}^i(x)\right)\right) \\
&+\frac{1}{8\pi}\left(\widehat{\vec{B}}^2(x)\delta_{ij} - \left(\widehat{B}^i(x)\widehat{B}^j(x) + \widehat{B}^j(x)\widehat{B}^i(x)\right)\right) \\
&= \frac{1}{8\pi}\begin{pmatrix}
\widehat{\vec{E}}^2(x) + \widehat{\vec{B}}^2(x) - 2\left(\widehat{E}_x^{\,2}(x) + \widehat{B}_x^{\,2}(x)\right) & -2\left(\widehat{E}_x(x)\widehat{E}_y(x) + \widehat{B}_x(x)\widehat{B}_y(x)\right) & -2\left(\widehat{E}_x(x)\widehat{E}_z(x) + \widehat{B}_x(x)\widehat{B}_z(x)\right) \\
-2\left(\widehat{E}_y(x)\widehat{E}_x(x) + \widehat{B}_y(x)\widehat{B}_x(x)\right) & \widehat{\vec{E}}2(x) + \widehat{\vec{B}}^2(x) - 2\left(\widehat{E}_y^{\,2}(x) + \widehat{B}_y^{\,2}(x)\right) & -2\left(\widehat{E}_y(x)\widehat{E}_z(x) + \widehat{B}_y(x)\widehat{B}_z(x)\right) \\
-2\left(\widehat{E}_z(x)\widehat{E}_x(x) + \widehat{B}_z(x)\widehat{B}_x(x)\right) & -2\left(\widehat{E}_z(x)\widehat{E}_y(x) + \widehat{B}_z(x)\widehat{B}_y(x)\right) & \widehat{\vec{E}}^2(x) + \widehat{\vec{B}}^2(x) - 2\left(\widehat{E}_z^{\,2}(x) + \widehat{B}_z^{\,2}(x)\right)
\end{pmatrix}
\end{aligned} \tag{1.37}$$

and $\widehat{\vec{L}}(x)$ is the Lorentz force density operator

$$\widehat{\vec{L}}(x) = \widehat{\vec{L}}_{\mathrm{e}}(x) + \sum_a \widehat{\vec{L}}_a(x) \tag{1.38}$$

$$\widehat{\vec{L}}_{\mathrm{e}}(x) = \widehat{\vec{E}}(x)\widehat{\rho}_{\mathrm{e}}(x) + \frac{1}{c}\widehat{\vec{j}}_{\mathrm{e}}(x) \times \widehat{\vec{B}}(x) \tag{1.39}$$

$$\widehat{\vec{L}}_a(x) = \widehat{\vec{E}}(x)\widehat{\rho}_a(x) + \frac{1}{c}\widehat{\vec{j}}_a(x) \times \widehat{\vec{B}}(x) \tag{1.40}$$

where $\widehat{\vec{L}}_{\mathrm{e}}(x)$ is the electronic Lorentz force density operator and $\widehat{\vec{L}}_a(x)$ is the Lorentz force density operator of a'th atomic nucleus. Note that $\widehat{\overleftrightarrow{\sigma}}(x)$ is symmetric

$$\widehat{\sigma}^{ij}(x) = \widehat{\sigma}^{ji}(x) \tag{1.41}$$

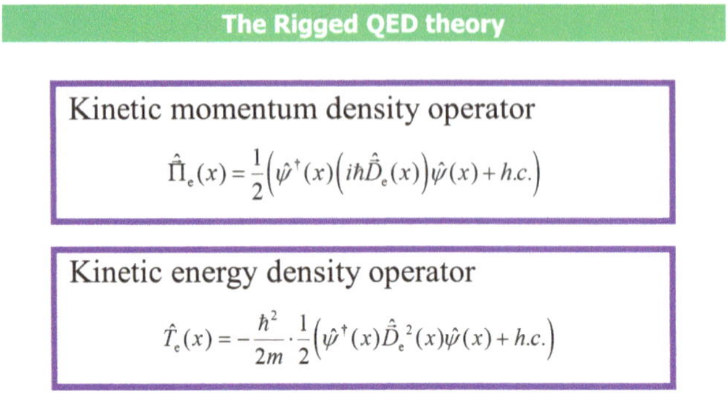

Fig. 1.21 Gauge-invariant kinetic momentum and energy density operator of electron

Secondly, the electronic kinetic momentum density operator (see Fig. 1.21)

$$\widehat{\vec{\Pi}}_e(x) = \frac{1}{2}\left(\widehat{\psi}^\dagger(x)\left(i\hbar\widehat{\vec{D}}_e(x)\right)\widehat{\psi}(x) + h.c.\right) \tag{1.42}$$

satisfies the equation of motion

$$\frac{\partial}{\partial t}\widehat{\vec{\Pi}}_e(x) = \widehat{\vec{L}}_e(x) + \widehat{\vec{\tau}}_e^{\Pi}(x) \tag{1.43}$$

The $\widehat{\vec{\tau}}_e^{\Pi}(x)$ is the electronic tension density operator given as the divergence of the electronic *internal self-stress tensor* density operator

$$\widehat{\vec{\tau}}_e^{\Pi}(x) = \mathrm{div}\widehat{\overleftrightarrow{\tau}}_e^{\Pi}(x), \quad \widehat{\tau}_e^{\Pi}k(x) = \partial_\ell \widehat{\tau}_e^{\Pi}k\ell(x) \tag{1.44}$$

$$\widehat{\tau}_e^{\Pi}\mu\nu(x) = \frac{c}{2}\left(\widehat{\overline{\psi}}(x)\gamma^\nu\left(-i\hbar\widehat{D}_e{}^\mu(x)\right)\widehat{\psi}(x) + h.c.\right) \tag{1.45}$$

It should be noted that $\widehat{\overleftrightarrow{\tau}}_e^{\Pi}(x)$ is Hermitean

$$\widehat{\overleftrightarrow{\tau}}_e^{\Pi\dagger}(x) = \widehat{\overleftrightarrow{\tau}}_e^{\Pi}(x) \tag{1.46}$$

Lastly, the kinetic momentum density operator $m_a\widehat{\vec{v}}_a(x)$ of atomic nucleus a satisfies the equation of motion

$$\frac{\partial}{\partial t}\left(m_a\widehat{\vec{v}}_a(x)\right) = \widehat{\vec{L}}_a(x) + \widehat{\vec{\tau}}_a^S(x) \tag{1.47}$$

The $\widehat{\vec{\tau}}_a^S(x)$ is the tension density operator given as the divergence of the nuclear *internal self-stress tensor* density operator

$$\hat{\vec{\tau}}_a^S(x) = \mathrm{div}\hat{\overleftrightarrow{\tau}}_a^S(x),\ \ \hat{\tau}_a^{Sk}(x) = \partial_\ell \hat{\tau}_a^{Sk\ell}(x) \tag{1.48}$$

$$\hat{\tau}_a^{Skl}(x) = \frac{\hbar^2}{4m_a}\left(\hat{\chi}_a^\dagger(x)\hat{D}_{ak}(x)\hat{D}_{al}(x)\hat{\chi}_a(x) - \left(\hat{D}_{ak}(x)\hat{\chi}_a(x)\right)^\dagger \cdot \hat{D}_{al}(x)\hat{\chi}_a(x) + h.c.\right) \tag{1.49}$$

It should be noted that the stress tensor density operator $\hat{\overleftrightarrow{\tau}}_a^S(x)$ is Hermitean and symmetric:

$$\hat{\overleftrightarrow{\tau}}_a^{S\dagger}(x) = \hat{\overleftrightarrow{\tau}}_a^S(x),\ \ \hat{\tau}_a^{Skl}(x) = \hat{\tau}_a^{Slk}(x) \tag{1.50}$$

As a whole, we obtain

$$\frac{\partial}{\partial t}\hat{\vec{\Pi}}(x) = \hat{\vec{L}}(x) + \hat{\vec{\tau}}(x) = \hat{\vec{L}}(x) + \mathrm{div}\hat{\overleftrightarrow{\tau}}(x) \tag{1.51}$$

$$\hat{\vec{\Pi}}(x) = \hat{\vec{\Pi}}_e(x) + \sum_a m_a \hat{\vec{v}}_a(x) \tag{1.52}$$

$$\hat{\vec{\tau}}(x) = \hat{\vec{\tau}}_e^\Pi(x) + \sum_a \hat{\vec{\tau}}_a^S(x) \tag{1.53}$$

$$\hat{\overleftrightarrow{\tau}}(x) = \hat{\overleftrightarrow{\tau}}_e^\Pi(x) + \sum_a \hat{\overleftrightarrow{\tau}}_a^S(x) \tag{1.54}$$

To sum up, we have

$$\frac{\partial}{\partial t}\left(\frac{1}{2}\left(\hat{\vec{G}}(x) + \hat{\vec{G}}^\dagger(x)\right) + \hat{\vec{\Pi}}(x)\right) = -\mathrm{div}\left(\hat{\overleftrightarrow{\sigma}}(x) - \hat{\overleftrightarrow{\tau}}(x)\right) \tag{1.55}$$

which is the momentum conservation law of the Rigged QED theory.

It should be noted the stress tensor itself is not defined uniquely (Heitler 1954; Tetrode 1928) since mathematically any tensor whose divergence is zero can be added to. Our stress tensor is defined in such a way that it appears in the equation of motion of $\hat{\vec{\Pi}}(x)$ as in Eqs. (1.51), (1.52), (1.53), (1.54), and (1.55).

1.2.7 Spin Torque of Electron

The electronic spin angular momentum density operator

$$\hat{\vec{s}}_e(x) = \frac{1}{2}\hbar\hat{\vec{\sigma}}_e(x),\ \ \hat{\vec{\sigma}}_e(x) = \hat{\psi}^\dagger(x)\vec{\sigma}\hat{\psi}(x) \tag{1.56}$$

satisfies the equation of motion

$$\frac{\partial}{\partial t}\widehat{\vec{s}}_{\mathrm{e}}(x) = \widehat{\vec{t}}_{\mathrm{e}}(x) + \widehat{\vec{\zeta}}_{\mathrm{e}}(x) \tag{1.57}$$

where $\widehat{\vec{t}}_{\mathrm{e}}(x)$ denotes the spin torque density operator defined as

$$\widehat{t}_{\mathrm{e}}^{k}(x) = -\varepsilon_{\ell nk}\widehat{\tau}_{\mathrm{e}}^{\Pi\ell n}(x) = -\varepsilon_{\ell nk}\widehat{\tau}_{\mathrm{e}}^{A\ell n}(x) \tag{1.58}$$

The $\widehat{\vec{\zeta}}_{\mathrm{e}}(x)$ denotes the zeta force density operator defined as

$$\widehat{\zeta}_{\mathrm{e}}^{k}(x) = -c\partial_{k}\left(\widehat{\overline{\psi}}(x)\gamma^{k}\frac{1}{2}\hbar\sigma^{k}\widehat{\psi}(x)\right); \text{no sum over } k \tag{1.59}$$

The alternative form using the gradient of the zeta potential $\phi_5(x)$ is obtained as follows:

$$\widehat{\zeta}_{\mathrm{e}}^{k}(x) = -\partial_{k}\widehat{\phi}_{5}(x) \tag{1.60}$$

$$\widehat{\phi}_{5}(x) = \frac{\hbar c}{2q_{\mathrm{e}}}\widehat{j}_{5}^{0}(x) = \frac{\hbar c^{2}}{2}\left(\widehat{N}_{R}(x) - \widehat{N}_{L}(x)\right). \tag{1.61}$$

where $\widehat{j}_{5}^{0}(x)$ denotes the zeroth component of the chiral current density operator

$$\widehat{j}_{5}^{\mu}(x) = cq_{\mathrm{e}}\widehat{\overline{\psi}}(x)\gamma^{\mu}\gamma_{5}\widehat{\psi}(x) \tag{1.62}$$

$$\gamma_{5} = i\gamma^{0}\gamma^{1}\gamma^{2}\gamma^{3} \tag{1.63}$$

with the chiral $\widehat{N}_{R,L}(x)$ components.

Thus, it is concluded that the electron spin torque is found to be counterbalanced by the chiral electron density. The geometrical nature of this relationship will further be discussed in Chap. 2.

1.2.8 Spin Vorticity of Electron

Since the vorticity $\mathrm{rot}\,\widehat{\vec{s}}_{\mathrm{e}}(x)$ is a solenoidal vector field, the spin $\widehat{\vec{s}}_{\mathrm{e}}(x)$ itself may be given by integration in the starlike domain using the rotation of torque $\widehat{\vec{t}}_{\mathrm{e}}(x)$ as the driving force

$$\begin{aligned}\mathrm{rot}\frac{\partial}{\partial t}\widehat{\vec{s}}_{\mathrm{e}}(\vec{r}) &= \mathrm{rot}\left(-\vec{r}\times\int_{0}^{1}\mathrm{rot}\frac{\partial}{\partial t}\widehat{\vec{s}}_{\mathrm{e}}(\lambda\vec{r})\lambda d\lambda\right)\\ &= \mathrm{rot}\left(-\vec{r}\times\int_{0}^{1}\mathrm{rot}\,\widehat{\vec{t}}_{\mathrm{e}}(\lambda\vec{r})\lambda d\lambda\right)\end{aligned} \tag{1.64}$$

where we have used that the zeta force $\widehat{\vec{\zeta}}_{\mathrm{e}}(x)$ is an irrotational conservative vector field as shown in Eq. (1.60). Moreover, it should be noted that

$$\mathrm{rot}\widehat{\vec{t}}_{\mathrm{e}}(x) = -2\mathrm{div}\widehat{\overleftrightarrow{\tau}}_{\mathrm{e}}^{A}(x) \tag{1.65}$$

and hence we obtain

$$\mathrm{rot}(\vec{s}_{\mathrm{e}}(t,\vec{r}) - \vec{s}_{\mathrm{e}}(t_0,\vec{r})) = \mathrm{rot}\left(2\vec{r}\times\int_{t_0}^{t}\left(\int_0^1 \mathrm{div}\overleftrightarrow{\tau}_{\mathrm{e}}^{A}(t',\lambda\vec{r})\lambda d\lambda\right)dt'\right) \tag{1.66}$$

with

$$\frac{\partial}{\partial t}\mathrm{rot}\widehat{\vec{s}}_{\mathrm{e}}(x) = -2\mathrm{div}\widehat{\overleftrightarrow{\tau}}_{\mathrm{e}}^{A}(x) \tag{1.67}$$

This is called the *quantum electron spin vorticity principle*: the time evolution of the electron spin $\widehat{\vec{s}}_{\mathrm{e}}(x)$ is driven by the antisymmetric component of the electronic stress tensor $\widehat{\overleftrightarrow{\tau}}_{\mathrm{e}}^{A}(x)$ through the vorticity $\mathrm{rot}\widehat{\vec{s}}_{\mathrm{e}}(x)$. If one half of Eq. (1.67) is added to Eq. (1.43), we get (see Fig. 1.22)

$$\frac{\partial}{\partial t}\left(\widehat{\vec{\Pi}}_{\mathrm{e}}(x) + \frac{1}{2}\mathrm{rot}\widehat{\vec{s}}_{\mathrm{e}}(x)\right) = \widehat{\vec{L}}_{\mathrm{e}}(x) + \mathrm{div}\widehat{\overleftrightarrow{\tau}}_{\mathrm{e}}^{S}(x) \tag{1.68}$$

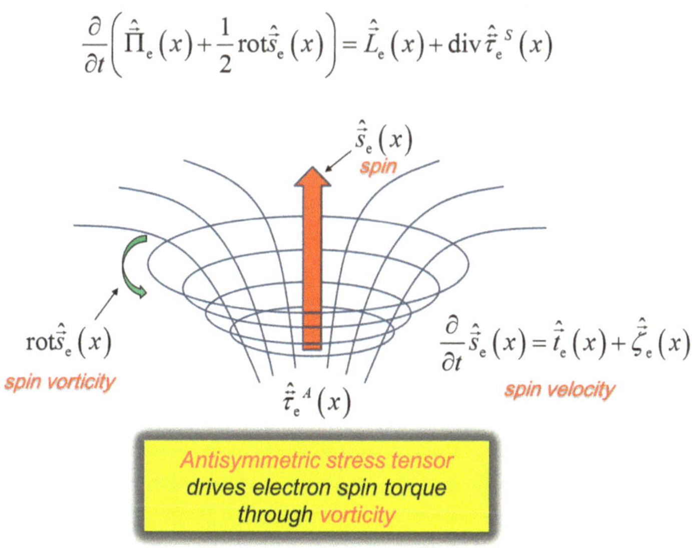

Fig. 1.22 Quantum electron spin vorticity principle

The applications to the other particles are also interesting. For example, for chiral spin-1/2 fermion with the non-Abelian gauge potential, analogous equation of motion of spin has been found (Tachibana 2010). Another example is the Majorana particle, which is neutral (Tachibana 2013).

We have proved that the spin vorticity of electron contributes to the kinetic momentum of electron. It raises a simple but "odd" question: what is momentum of electron spin? How odd this question is should be obvious since electron is considered a point particle, and spin is its internal degree of freedom and then spin is considered to have nothing to do with momentum. In the next chapter, we shall resolve this question.

1.2.9 Angular Momentum of QED

The angular momentum density operator $\hat{\vec{u}}(x)$ of electromagnetic field defined as

$$\hat{\vec{u}}(x) = \vec{r} \times \hat{\vec{G}}(x) \tag{1.69}$$

satisfies the equation of motion

$$\begin{aligned}\frac{\partial}{\partial t}\frac{1}{2}\left(\hat{\vec{u}}(x) + \hat{\vec{u}}^{\dagger}(x)\right) &= -\vec{r} \times \left(\frac{1}{2}\left(\hat{\vec{L}}(x) + \hat{\vec{L}}^{\dagger}(x)\right) + \mathrm{div}\hat{\overleftrightarrow{\sigma}}(x)\right) \\ &= -\vec{r} \times \frac{1}{2}\left(\hat{\vec{L}}(x) + \hat{\vec{L}}^{\dagger}(x)\right) - \mathrm{div}\left(\vec{r} \times \hat{\overleftrightarrow{\sigma}}(x)\right)\end{aligned} \tag{1.70}$$

The electronic orbital angular momentum density operator $\hat{\vec{\ell}}_{\mathrm{e}}(x)$ defined as

$$\hat{\vec{\ell}}_e(x) = \vec{r} \times \hat{\vec{\Pi}}_e(x) \tag{1.71}$$

satisfies the equation of motion

$$\frac{\partial}{\partial t}\hat{\vec{\ell}}_{\mathrm{e}}(x) = \vec{r} \times \left(\hat{\vec{L}}_{\mathrm{e}}(x) + \mathrm{div}\hat{\overleftrightarrow{\tau}}_{\mathrm{e}}^{\Pi}(x)\right) = \vec{r} \times \hat{\vec{L}}_{\mathrm{e}}(x) + \mathrm{div}\left(\vec{r} \times \hat{\overleftrightarrow{\tau}}_{\mathrm{e}}^{\Pi}(x)\right) \tag{1.72}$$

Sum of Eqs. (1.57) and (1.72) leads to

$$\frac{\partial}{\partial t}\left(\hat{\vec{\ell}}_{\mathrm{e}}(x) + \hat{\vec{s}}_{\mathrm{e}}(x)\right) = \vec{r} \times \hat{\vec{L}}_{\mathrm{e}}(x) + \mathrm{div}\left(\vec{r} \times \hat{\overleftrightarrow{\tau}}_{\mathrm{e}}^{\Pi}(x)\right) + \hat{\vec{t}}_{\mathrm{e}}(x) + \hat{\vec{\zeta}}_{\mathrm{e}}(x) \tag{1.73}$$

The a'th nuclear orbital angular momentum density operator $\hat{\vec{\ell}}_a(x)$ defined as

$$\hat{\vec{\ell}}_a(x) = \vec{r} \times m_a \hat{\vec{v}}_a(x) \tag{1.74}$$

satisfies the equation of motion:

$$\frac{\partial}{\partial t}\hat{\vec{\ell}}_a(x) = \vec{r} \times \hat{\vec{L}}_a(x) + \mathrm{div}\left(\vec{r} \times \hat{\overleftrightarrow{\tau}}{}^S_a(x)\right) \tag{1.75}$$

To sum up, we have

$$\frac{\partial}{\partial t}\left(\hat{\vec{\ell}}_e(x) + \hat{\vec{s}}_e(x) + \sum_a \hat{\vec{\ell}}_a(x)\right) = \vec{r} \times \hat{\vec{L}}(x) + \mathrm{div}\left(\vec{r} \times \hat{\overleftrightarrow{\tau}}(x)\right) + \hat{\vec{t}}_e(x) + \hat{\vec{\zeta}}_e(x) \tag{1.76}$$

If the time derivative of $\hat{\vec{u}}(x)$ from Eq. (1.70) is further added to, we finally obtain

$$\begin{aligned}&\frac{\partial}{\partial t}\left(\frac{1}{2}\left(\hat{\vec{u}}(x) + \hat{\vec{u}}^\dagger(x)\right) + \hat{\vec{\ell}}_e(x) + \hat{\vec{s}}_e(x) + \sum_a \hat{\vec{\ell}}_a(x)\right)\\ &= -\vec{r} \times \mathrm{div}\left(\hat{\overleftrightarrow{\sigma}}(x) - \hat{\overleftrightarrow{\tau}}(x)\right) + \hat{\vec{t}}_e(x) + \hat{\vec{\zeta}}_e(x)\end{aligned} \tag{1.77}$$

If we use the electron spin vorticity principle in the form

$$\vec{r} \times \frac{\partial}{\partial t}\left(\frac{1}{2}\mathrm{rot}\hat{\vec{s}}_e(x)\right) = -\vec{r} \times \left(\mathrm{div}\hat{\overleftrightarrow{\tau}}{}^A_e(x)\right) = -\mathrm{div}\left(\vec{r} \times \hat{\overleftrightarrow{\tau}}{}^A_e(x)\right) \tag{1.78}$$

then we get

$$\frac{\partial}{\partial t}\left(\vec{r} \times \frac{1}{2}\mathrm{rot}\hat{\vec{s}}_e(x) - \hat{\vec{s}}_e(x)\right) = -\mathrm{div}\left(\vec{r} \times \hat{\overleftrightarrow{\tau}}{}^A_e(x)\right) - \left(\hat{\vec{t}}_e(x) + \hat{\vec{\zeta}}_e(x)\right) \tag{1.79}$$

so that we arrive at

$$\begin{aligned}&\frac{\partial}{\partial t}\left(\frac{1}{2}\left(\hat{\vec{u}}(x) + \hat{\vec{u}}^\dagger(x)\right) + \hat{\vec{\ell}}_e(x) + \vec{r} \times \frac{1}{2}\mathrm{rot}\hat{\vec{s}}_e(x) + \sum_a \hat{\vec{\ell}}_a(x)\right)\\ &= -\mathrm{div}\left(\vec{r} \times \left(\hat{\overleftrightarrow{\sigma}}(x) - \hat{\overleftrightarrow{\tau}}{}^S(x)\right)\right)\end{aligned} \tag{1.80}$$

which is the angular momentum conservation law of the Rigged QED theory.

Mechanical measurement of the angular momentum of light has been performed experimentally (see Fig. 1.23).

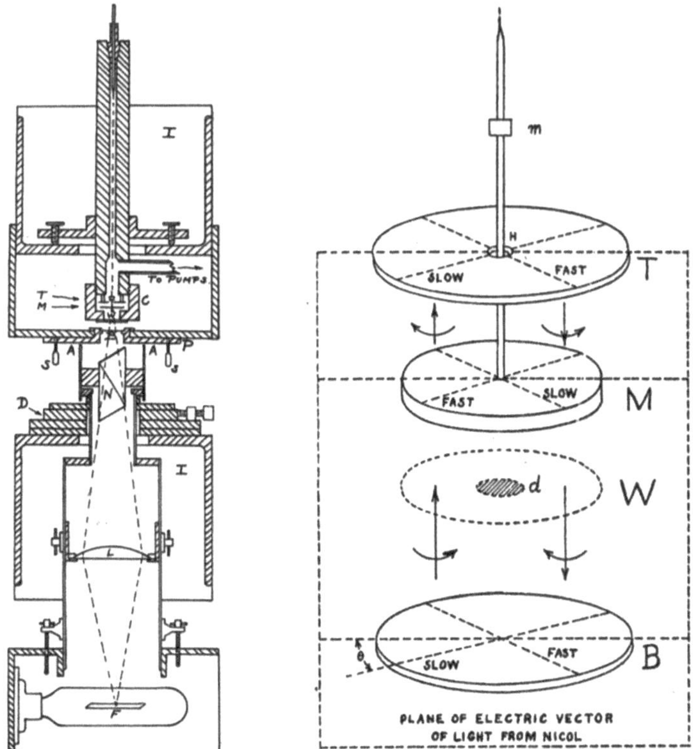

FIG. 1. Diagram of apparatus. FIG. 3. Wave plate arrangement.

Fig. 1.23 Mechanical measurement of the angular momentum of light (Reproduced from Beth 1935)

1.3 Phenomenology of the Rigged QED Theory

1.3.1 Energy Density

The QED Hamiltonian density operator $\widehat{H}_{\mathrm{QED}}(x)$ is composed of the Hamiltonian density operator of the electromagnetic field $\widehat{H}_{\mathrm{EM}}(x)$ and the Dirac electronic Hamiltonian density operator $\widehat{H}_{\mathrm{Dirac}}(x)$ interacting with the electromagnetic field

$$\widehat{H}_{\mathrm{QED}}(x) = \widehat{H}_{\mathrm{EM}}(x) + \widehat{H}_{\mathrm{Dirac}}(x) \tag{1.81}$$

$$\widehat{H}_{\mathrm{EM}}(x) = \widehat{H}_{\gamma}(x) - \widehat{A}_0(x)\widehat{\rho}_{\mathrm{e}}(x) \tag{1.82}$$

$$\widehat{H}_{\mathrm{Dirac}}(x) = \widehat{M}_{\mathrm{e}}(x) + \widehat{A}_0(x)\widehat{\rho}_{\mathrm{e}}(x) \tag{1.83}$$

where $\widehat{H}_\gamma(x)$ is the electromagnetic field energy density operator and $\widehat{M}_\mathrm{e}(x)$ is the electronic mass density operator:

$$\widehat{H}_\gamma(x) = \frac{1}{8\pi}\left(\widehat{\vec{E}}^2(x) + \widehat{\vec{B}}^2(x)\right) \tag{1.84}$$

$$\widehat{M}_\mathrm{e}(x) = c\widehat{\overline{\psi}}(x)\left(-i\hbar\gamma^k\widehat{D}_{\mathrm{e}k}(x) + m_\mathrm{e}c\right)\widehat{\psi}(x) \tag{1.85}$$

The electronic mass density operator $\widehat{M}_\mathrm{e}(x)$ may be written as the energy density operator of electron $\widehat{H}_\mathrm{e}(x)$ as follows:

$$\widehat{M}_\mathrm{e}(x) = \widehat{H}_\mathrm{e}(x) \tag{1.86}$$

Thus, the $\widehat{H}_\mathrm{QED}(x)$ reduces to (see Fig. 1.24)

$$\widehat{H}_\mathrm{QED}(x) = \widehat{H}_\gamma(x) + \widehat{H}_\mathrm{e}(x) \tag{1.87}$$

The Rigged QED Hamiltonian density operator denoted as $\widehat{H}_\mathrm{Rigged\ QED}(x)$ is derived as follows:

$$\widehat{H}_\mathrm{Rigged\ QED}(x) = \widehat{H}_\mathrm{QED}(x) + \widehat{H}_\mathrm{atom}(x) \tag{1.88}$$

where the energy density operator $\widehat{H}_\mathrm{atom}(x)$ of atomic nuclei interacting through the electromagnetic field and the electron field is added to $\widehat{H}_\mathrm{QED}(x)$. The $\widehat{H}_\mathrm{atom}(x)$ is purely the kinetic energy density operator of atomic nuclei

$$\widehat{H}_\mathrm{atom}(x) = \sum_a \widehat{T}_a(x) \tag{1.89}$$

$$\widehat{T}_a(x) = -\frac{\hbar^2}{2m_a}\cdot\frac{1}{2}\left(\widehat{\chi}_a^\dagger(x)\widehat{\vec{D}}_a{}^2(x)\widehat{\chi}_a(x) + h.c.\right) \tag{1.90}$$

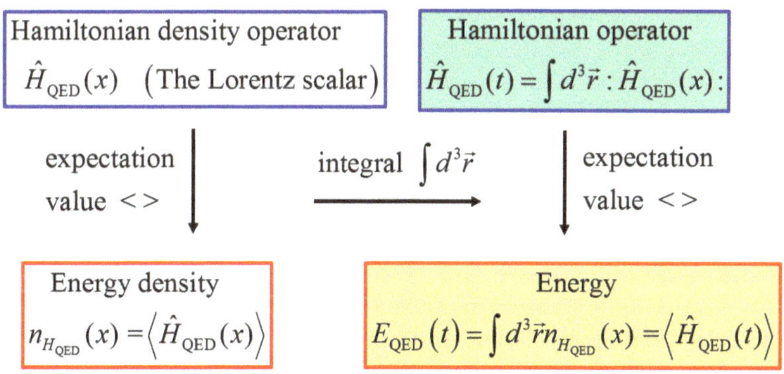

Fig. 1.24 Energy density concept in the field theory

The energy flow is found to be

$$\frac{\partial}{\partial t}\widehat{H}_{\gamma}(x) = -c^2 \mathrm{div}\frac{1}{2}\left(\widehat{\vec{G}}(x) + \widehat{\vec{G}}^{\dagger}(x)\right) - \frac{1}{2}\left(\widehat{\vec{E}}(x)\bullet\widehat{\vec{j}}(x) + \widehat{\vec{j}}(x)\bullet\widehat{\vec{E}}(x)\right) \tag{1.91}$$

$$\frac{\partial}{\partial t}\widehat{H}_{\mathrm{e}}(x) = -c^2 \mathrm{div}\,\widehat{\vec{\Pi}}_{\mathrm{e}}(x) + \frac{1}{2}\left(\widehat{\vec{E}}(x)\bullet\widehat{\vec{j}}_{\mathrm{e}}(x) + \widehat{\vec{j}}_{\mathrm{e}}(x)\bullet\widehat{\vec{E}}(x)\right) \tag{1.92}$$

$$\frac{\partial}{\partial t}\widehat{H}_{\mathrm{atom}}(x) = -\mathrm{div}\sum_a \widehat{\vec{s}}_a(x) + \frac{1}{2}\left(\widehat{\vec{E}}(x)\bullet\sum_a \widehat{\vec{j}}_a(x) + \sum_a \widehat{\vec{j}}_a(x)\bullet\widehat{\vec{E}}(x)\right) \tag{1.93}$$

with

$$\widehat{s}_a^k(x) = \frac{1}{2i\hbar}\left(\frac{\hbar^2}{2m_a}\right)\left(-\widehat{\chi}_a^{\dagger}(x)\widehat{D}_{ak}(x)\widehat{\vec{D}}_a^2(x)\widehat{\chi}_a(x) + \left(\widehat{D}_{ak}(x)\widehat{\chi}_a(x)\right)^{\dagger}\widehat{\vec{D}}_a^2(x)\widehat{\chi}_a(x) - h.c.\right) \tag{1.94}$$

leading to

$$\frac{\partial}{\partial t}\widehat{H}_{\text{Rigged QED}}(x) = -\mathrm{div}\left(c^2\frac{1}{2}\left(\widehat{\vec{G}}(x) + \widehat{\vec{G}}^{\dagger}(x)\right) + c^2\widehat{\vec{\Pi}}_{\mathrm{e}}(x) + \sum_a \widehat{\vec{s}}_a(x)\right) \tag{1.95}$$

which is the energy conservation law of the Rigged QED theory (see Fig. 1.25 with Eqs. (1.55) and (1.80)).

Energy conservation law

$$\frac{\partial}{\partial t}\hat{H}_{\text{Rigged QED}}(x) = -\mathrm{div}\left(c^2\frac{1}{2}\left(\hat{\vec{G}}(x) + \hat{\vec{G}}^{\dagger}(x)\right) + c^2\hat{\vec{\Pi}}_{\mathrm{e}}(x) + \sum_a \hat{\vec{s}}_a(x)\right)$$

Momentum conservation law

$$\frac{\partial}{\partial t}\left(\frac{1}{2}\left(\hat{\vec{G}}(x) + \hat{\vec{G}}^{\dagger}(x)\right) + \hat{\vec{\Pi}}(x)\right) = -\mathrm{div}\left(\hat{\vec{\sigma}}(x) - \hat{\vec{\tau}}(x)\right)$$

Angular momentum conservation law

$$\frac{\partial}{\partial t}\left(\frac{1}{2}\left(\hat{\vec{u}}(x) + \hat{\vec{u}}^{\dagger}(x)\right) + \hat{\vec{\ell}}_{\mathrm{e}}(x) + \vec{r}\times\frac{1}{2}\mathrm{rot}\,\hat{\vec{s}}_{\mathrm{e}}(x) + \sum_a \hat{\vec{\ell}}_a(x)\right)$$
$$= -\mathrm{div}\left(\vec{r}\times\left(\hat{\vec{\sigma}}(x) - \hat{\vec{\tau}}^{S}(x)\right)\right)$$

Fig. 1.25 Energy, momentum, and angular momentum conservation laws of the Rigged QED theory

It should be noted that the application of the Noether theorem associated with the canonical quantization is the textbook approach to derive the conservation laws (Weinberg 1995).

We shall further use the virial theorem (Landau and Lifshitz 1975) for the finite-system energetics of the Rigged QED theory in Sect. 2.2, Chap. 2.

1.3.2 Electromagnetic Energy Density in Magnetodielectric Media

In the Rigged QED theory, the phenomenological interaction of a system A and its environment background medium M is tractable using regional charge and current densities. For phenomenological force concepts in magnetodielectric medium such as chemical reaction systems in condensed phase, we may usually rely on a classical analogy of parallel-plate capacitor filled with a dielectric (see Fig. 1.26). Nuclear magnetic currents for nuclear spin (Itzykson and Zuber 1980) can be treated as if they were within M.

The corresponding gauge potentials are the regional integrals of the charge and transversal current densities, defined as follows (Tachibana 2010)

$$\widehat{A}_{0_{\mathrm{A}}}(ct, \vec{r}) = \int_{\mathrm{A}} d^3\vec{s} \frac{\widehat{\rho}(ct, \vec{s})}{|\vec{r} - \vec{s}|} \tag{1.96}$$

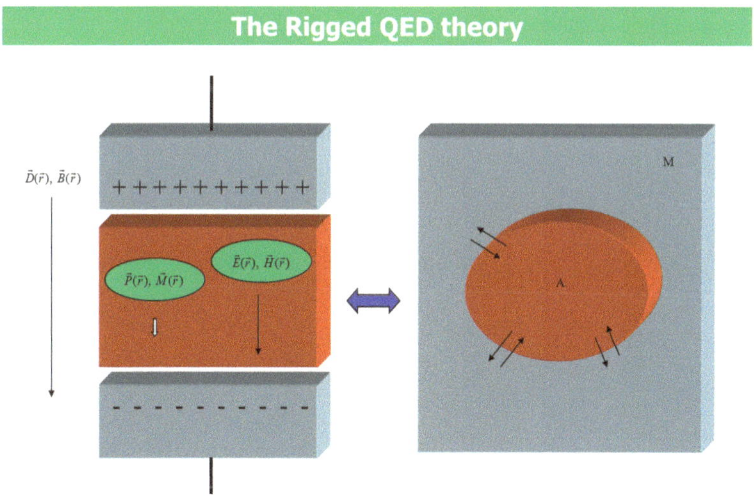

Fig. 1.26 Parallel-plate capacitor filled with a dielectric: a phenomenological model of a chemical reaction system A embedded in an environmental background medium M

$$\widehat{A}_{0_M}(ct, \vec{r}) = \int_M d^3\vec{s}\frac{\widehat{\rho}(ct, \vec{s})}{|\vec{r} - \vec{s}|} \tag{1.97}$$

and

$$\widehat{\vec{A}}_A(ct, \vec{r}) = \frac{1}{c}\int_A d^3\vec{s}\frac{\widehat{\vec{j}}_T(cu, \vec{s})}{|\vec{r} - \vec{s}|} \tag{1.98}$$

$$\widehat{\vec{A}}_M(ct, \vec{r}) = \frac{1}{c}\int_M d^3\vec{s}\frac{\widehat{\vec{j}}_T(cu, \vec{s})}{|\vec{r} - \vec{s}|} \tag{1.99}$$

where the subscript A or M of the integral sign denotes the regional integrals confined to the region A or M, respectively, and where $u = t - \frac{|\vec{r}-\vec{s}|}{c}$ (see Fig. 1.27).

Since the regions A and M altogether span the whole space, we have

$$\widehat{A}_0(x) = \widehat{A}_{0_A}(x) + \widehat{A}_{0_M}(x) \tag{1.100}$$

$$\widehat{\vec{A}}(x) = \widehat{\vec{A}}_{\text{radiation}}(x) + \widehat{\vec{A}}_A(x) + \widehat{\vec{A}}_M(x) \tag{1.101}$$

where the radiation gauge potential satisfies

$$\left(\frac{1}{c^2}\frac{\partial^2}{\partial t^2} - \Delta\right)\widehat{\vec{A}}_{\text{radiation}}(\vec{r}) = 0, \quad div\widehat{\vec{A}}_{\text{radiation}}(\vec{r}) = 0 \tag{1.102}$$

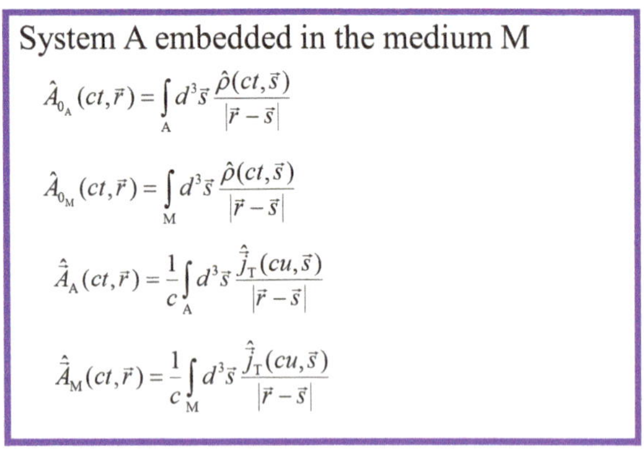

Fig. 1.27 Gauge potentials in an environmental background medium M

This is given as

$$\widehat{A}_{\text{radiation}}{}^{\mu}(x) = \frac{1}{\sqrt{(2\pi\hbar)^3}} \sum_{\sigma=\pm 1} \int \frac{d^3\vec{p}}{\sqrt{2p^0}} \sqrt{4\pi\hbar^2 c}\left(e^{\mu}(\vec{p},\sigma)\widehat{a}(\vec{p},\sigma)e^{-ix\cdot p/\hbar} + e^{\mu *}(\vec{p},\sigma)\widehat{a}^{\dagger}(\vec{p},\sigma)e^{ix\cdot p/\hbar}\right) \tag{1.103}$$

The polarization vector is

$$e^{\mu}(\vec{p},\sigma)e^{-i\sigma\theta(L(p;k),p;k)} = R^{\mu}{}_{\nu}\left(\widehat{\vec{p}}\right)e^{\nu}\left(\vec{k},\sigma\right) = \frac{1}{\sqrt{2}}\begin{pmatrix} 0 \\ \cos\phi\cos\theta \mp i\sin\phi \\ \sin\phi\cos\theta \pm i\cos\phi \\ -\sin\theta \end{pmatrix} \tag{1.104}$$

$$e^{0}(\vec{p},\sigma) = 0 \tag{1.105}$$

$$p^{k}e_{k}(\vec{p},\sigma) = 0 \tag{1.106}$$

$$\sum_{\sigma=\pm 1} e^{i}(\vec{p},\sigma)e^{j*}(\vec{p},\sigma) = -\eta^{ij} + \frac{p^i p^j}{-\vec{p}^2} \tag{1.107}$$

The electric field $\widehat{\vec{E}}(x)$ is decomposed into the electric displacement $\widehat{\vec{D}}(x)$ of the medium M and the polarization $\widehat{\vec{P}}(x)$ of the system A, defined, respectively, as

$$\widehat{\vec{D}}(x) = -\text{grad}\widehat{A}_{0_{\text{M}}}(x) - \frac{1}{c}\frac{\partial}{\partial t}\widehat{\vec{A}}_{\text{M}}(x) \tag{1.108}$$

$$\widehat{\vec{P}}(x) = \frac{1}{4\pi}\text{grad}\widehat{A}_{0_{\text{A}}}(x) + \frac{1}{4\pi c}\frac{\partial}{\partial t}\widehat{\vec{A}}_{\text{A}}(x) \tag{1.109}$$

so that we have (see Fig. 1.28)

Electric field

$$\hat{\vec{E}}(x) = -\text{grad}\hat{A}_0(x) - \frac{1}{c}\frac{\partial}{\partial t}\hat{\vec{A}}(x) = \hat{\vec{D}}(x) - 4\pi\hat{\vec{P}}(x) - \frac{1}{c}\frac{\partial}{\partial t}\hat{\vec{A}}_{\text{radiation}}(x)$$

$$\hat{\vec{D}}(x) = -\text{grad}\hat{A}_{0_{\text{M}}}(x) - \frac{1}{c}\frac{\partial}{\partial t}\hat{\vec{A}}_{\text{M}}(x), \quad \hat{\vec{P}}(x) = \frac{1}{4\pi}\text{grad}\hat{A}_{0_{\text{A}}}(x) + \frac{1}{4\pi c}\frac{\partial}{\partial t}\hat{\vec{A}}_{\text{A}}(x)$$

Magnetic field

$$\hat{\vec{B}}(x) = \text{rot}\hat{\vec{A}}(x) = \hat{\vec{H}}(x) + 4\pi\hat{\vec{M}}(x)$$

$$\hat{\vec{H}}(x) = \hat{\vec{H}}_{\text{M}}(x) + \text{rot}\hat{\vec{A}}_{\text{radiation}}(x), \quad \hat{\vec{M}}(x) = \frac{1}{4\pi}\text{rot}\hat{\vec{A}}_{\text{A}}(x)$$

Fig. 1.28 Electromagnetic fields in the system A and an environmental background medium M

$$\widehat{\vec{E}}(x) = -\mathrm{grad}\widehat{A}_0(x) - \frac{1}{c}\frac{\partial}{\partial t}\widehat{\vec{A}}(x) = \widehat{\vec{D}}(x) - 4\pi\widehat{\vec{P}}(x) - \frac{1}{c}\frac{\partial}{\partial t}\widehat{\vec{A}}_{\mathrm{radiation}}(x) \quad (1.110)$$

Likewise, let the magnetic field $\widehat{\vec{H}}(x)$ of the medium M and the magnetization $\widehat{\vec{M}}(x)$ of the system A be defined, respectively, as

$$\widehat{\vec{H}}_{\mathrm{M}}(x) = \mathrm{rot}\widehat{\vec{A}}_{\mathrm{M}}(x) \quad (1.111)$$

$$\widehat{\vec{M}}(x) = \frac{1}{4\pi}\mathrm{rot}\widehat{\vec{A}}_{\mathrm{A}}(x) \quad (1.112)$$

then we have (see Fig. 1.17)

$$\widehat{\vec{B}}(x) = \mathrm{rot}\widehat{\vec{A}}(x) \quad (1.113)$$

$$\mathrm{rot}\widehat{\vec{A}}(x) = \widehat{\vec{H}}(x) + 4\pi\widehat{\vec{M}}(x) \quad (1.114)$$

$$\widehat{\vec{H}}(x) = \widehat{\vec{H}}_{\mathrm{M}}(x) + \mathrm{rot}\widehat{\vec{A}}_{\mathrm{radiation}}(x) \quad (1.115)$$

The regional charge densities are then represented, respectively, as

$$\widehat{\rho}_{\mathrm{A}}(x) = -\frac{1}{4\pi}\Delta\widehat{A}_{0_{\mathrm{A}}}(x) \quad (1.116)$$

$$\widehat{\rho}_{\mathrm{M}}(x) = -\frac{1}{4\pi}\Delta\widehat{A}_{0_{\mathrm{M}}}(x) \quad (1.117)$$

and hence

$$\widehat{\rho}(x) = \widehat{\rho}_{\mathrm{A}}(x) + \widehat{\rho}_{\mathrm{M}}(x) \quad (1.118)$$

The regional charge current densities are represented as

$$\widehat{\vec{j}}_{\mathrm{A}}(x) = \frac{c}{4\pi}\left(\frac{1}{c}\mathrm{grad}\frac{\partial}{\partial t}\widehat{A}_{0_{\mathrm{A}}}(x) + \Box\widehat{\vec{A}}_{\mathrm{A}}(x)\right) = \frac{\partial}{\partial t}\widehat{\vec{P}}(x) + c\,\mathrm{rot}\widehat{\vec{M}}(x) \quad (1.119)$$

$$\widehat{\vec{j}}_{\mathrm{M}}(x) = \frac{c}{4\pi}\left(\frac{1}{c}\mathrm{grad}\frac{\partial}{\partial t}\widehat{A}_{0_{\mathrm{M}}}(x) + \Box\widehat{\vec{A}}_{\mathrm{M}}(x)\right) \quad (1.120)$$

$$\begin{aligned}\widehat{\vec{j}}(x) &= \widehat{\vec{j}}_{\mathrm{A}}(x) + \widehat{\vec{j}}_{\mathrm{M}}(x)\\ &= \frac{\partial}{\partial t}\widehat{\vec{P}}(x) + c\,\mathrm{rot}\widehat{\vec{M}}(x) + \widehat{\vec{j}}_{\mathrm{M}}(x)\end{aligned} \quad (1.121)$$

The regional decomposition of the longitudinal and transversal components of the current densities are represented as follows:

$$\widehat{\vec{j}}(x) = \widehat{\vec{j}}_{\mathrm{L}}(x) + \widehat{\vec{j}}_{\mathrm{T}}(x) \tag{1.122}$$

with

$$\widehat{\vec{j}}_{\mathrm{L}}(x) = \widehat{\vec{j}}_{\mathrm{L_A}}(x) + \widehat{\vec{j}}_{\mathrm{L_M}}(x) \tag{1.123}$$

$$\widehat{\vec{j}}_{\mathrm{T}}(x) = \widehat{\vec{j}}_{\mathrm{T_A}}(x) + \widehat{\vec{j}}_{\mathrm{T_M}}(x) \tag{1.124}$$

where

$$\widehat{\vec{j}}_{\mathrm{L_A}}(x) = \frac{c}{4\pi} \cdot \frac{1}{c} \mathrm{grad} \frac{\partial}{\partial t} \widehat{A}_{0_{\mathrm{A}}}(x) \tag{1.125}$$

$$\widehat{\vec{j}}_{\mathrm{L_M}}(x) = \frac{c}{4\pi} \cdot \frac{1}{c} \mathrm{grad} \frac{\partial}{\partial t} \widehat{A}_{\mathrm{M}}(x) \tag{1.126}$$

$$\widehat{\vec{j}}_{\mathrm{T_A}}(x) = \frac{c}{4\pi} \cdot \Box \widehat{\vec{A}}_{\mathrm{A}}(x) \tag{1.127}$$

$$\widehat{\vec{j}}_{\mathrm{T_M}}(x) = \frac{c}{4\pi} \cdot \Box \widehat{\vec{A}}_{\mathrm{M}}(x) \tag{1.128}$$

We have the alternative forms as

$$\widehat{\vec{j}}_{\mathrm{A}}(x) = \widehat{\vec{j}}_{\mathrm{L_A}}(x) + \widehat{\vec{j}}_{\mathrm{T_A}}(x) \tag{1.129}$$

$$\widehat{\vec{j}}_{\mathrm{M}}(x) = \widehat{\vec{j}}_{\mathrm{L_M}}(x) + \widehat{\vec{j}}_{\mathrm{T_M}}(x) \tag{1.130}$$

The linear response properties of the system A under the interaction with the environment medium M may formally be represented with obvious notation as follows

$$\begin{aligned}\widehat{\vec{P}}(x) &= \widehat{\overleftrightarrow{\alpha}}(x)\left(\widehat{\vec{D}}(x) - \frac{1}{c}\frac{\partial}{\partial t}\widehat{\vec{A}}_{\mathrm{radiation}}(x)\right)\\ &= \widehat{\overleftrightarrow{\chi}}_e(x)\widehat{\vec{E}}(x)\end{aligned} \tag{1.131}$$

$$\widehat{\vec{M}}(x) = \widehat{\overleftrightarrow{\chi}}_m(x)\widehat{\vec{H}}(x) \tag{1.132}$$

$$\begin{aligned}&\widehat{\vec{D}}(x) - \frac{1}{c}\frac{\partial}{\partial t}\widehat{\vec{A}}_{\mathrm{radiation}}(x) = \left(1 + 4\pi\widehat{\overleftrightarrow{\chi}}_e(x)\right)\widehat{\vec{E}}(x)\\ &= \frac{1}{1 - 4\pi\widehat{\overleftrightarrow{\alpha}}}(x)\widehat{\vec{E}}(x)\\ &= \widehat{\overleftrightarrow{\varepsilon}}(x)\widehat{\vec{E}}(x)\end{aligned} \tag{1.133}$$

$$\begin{aligned}\widehat{\vec{B}}(x) &= \left(1 + 4\pi\widehat{\overleftrightarrow{\chi}}_m(x)\right)\widehat{\vec{H}}(x) \\ &= \widehat{\overleftrightarrow{\mu}}(x)\widehat{\vec{H}}(x)\end{aligned} \tag{1.134}$$

and

$$\begin{aligned}\widehat{\vec{j}}(x) &= \widehat{\overleftrightarrow{\sigma}}_{\text{ext}}(x)\left(\widehat{\vec{D}}(x) - \frac{1}{c}\frac{\partial}{\partial t}\widehat{\vec{A}}_{\text{radiation}}(x)\right) \\ &= \widehat{\overleftrightarrow{\sigma}}_{\text{ext}}(x)\widehat{\overleftrightarrow{\varepsilon}}(x)\widehat{\vec{E}}(x) \\ &= \widehat{\overleftrightarrow{\sigma}}_{\text{int}}(x)\widehat{\vec{E}}(x)\end{aligned} \tag{1.135}$$

Photon deflection is realized by the index of refraction

$$\widehat{\overleftrightarrow{n}}(x) = \sqrt{\widehat{\overleftrightarrow{\mu}}(x)\widehat{\overleftrightarrow{\varepsilon}}(x)} \tag{1.136}$$

Chirality of matter affects the helicity of photon (photon spin) $\widehat{\vec{S}}(x)$ in the wave zone (see Fig. 1.29)

$$\widehat{\vec{S}}(x) = \frac{1}{4\pi c}\widehat{\vec{E}}(x) \times \widehat{\vec{A}}(x) = \widehat{\vec{S}}_{\text{radiation}}(x) + \widehat{\vec{S}}_{\text{matter}}(x) \tag{1.137}$$

$$\widehat{\vec{S}}_{\text{radiation}}(x) = \frac{1}{4\pi c}\widehat{\vec{E}}_{\text{radiation}}(x) \times \widehat{\vec{A}}_{\text{radiation}}(x),$$

$$\widehat{\vec{E}}_{\text{radiation}}(x) = -\frac{1}{c}\frac{\partial}{\partial t}\widehat{\vec{A}}_{\text{radiation}}(x) \tag{1.138}$$

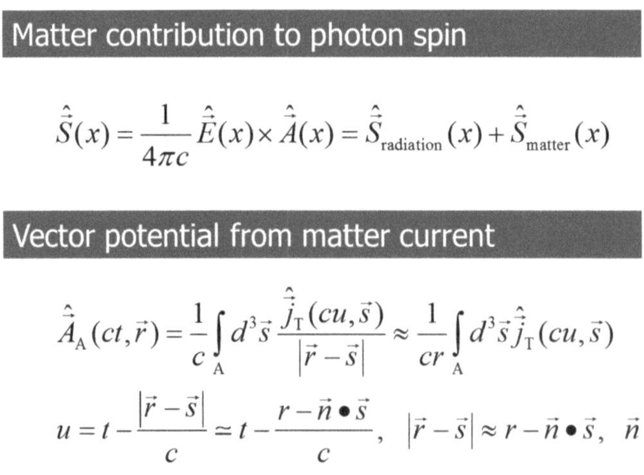

Fig. 1.29 Photon spin originated from matter current in the wave zone

If atoms and molecules are irradiated, then the electrons may be affected by torque leading to the imbalance in between spin torque and zeta force. The back reaction may also affect the torque on photon, leading to the circular dichroism, the Kerr effect, or the Faraday effect. Forbidden processes may of course occur due to forbidden symmetry of the ket vectors.

1.3.3 Effective Charge Number of Electromigration

Electromigration is the phenomena of nuclear current induced by electric current in condensed phase (Lodder and Dekker 1998). The nuclei accept diffusive force from the surrounding medium over and above the Lorentz force (Bosvieux and Friedel 1962). In our model, the tension is the origin of the medium effects.

The linear response of the force defines the effective charge number tensor density operator $\hat{\overleftrightarrow{Z}}^{*}_{\alpha}(x)$ of α'th charged particle as

$$\hat{\overleftrightarrow{Z}}^{*}_{\alpha}(x)e\left(\hat{\vec{D}}(x)-\frac{1}{c}\frac{\partial}{\partial t}\hat{\vec{A}}_{\text{radiation}}(x)\right)\hat{N}_{\alpha}(x)+\frac{1}{c}\hat{\vec{j}}_{\alpha}(x)\times\hat{\vec{B}}(x)=\hat{\vec{L}}_{\alpha}(x)+\hat{\vec{\tau}}^{S}_{\alpha}(x) \tag{1.139}$$

Since the right-hand side of this equation is

$$\hat{\vec{L}}_{\alpha}(x)+\hat{\vec{\tau}}^{S}_{\alpha}(x)=\hat{\vec{E}}(x)\hat{\rho}_{\alpha}(x)+\frac{1}{c}\hat{\vec{j}}_{\alpha}(x)\times\hat{\vec{B}}(x)+\hat{\vec{\tau}}^{S}_{\alpha}(x) \tag{1.140}$$

we then conclude

$$\begin{aligned}
&\hat{\overleftrightarrow{Z}}^{*}_{\alpha}(x)e\left(\hat{\vec{D}}(x)-\frac{1}{c}\frac{\partial}{\partial t}\hat{\vec{A}}_{\text{radiation}}(x)\right)\hat{N}_{\alpha}(x)=\hat{\vec{E}}(x)\hat{\rho}_{\alpha}(x)+\hat{\vec{\tau}}^{S}_{\alpha}(x)\\
&=\left(\hat{\vec{D}}(x)-4\pi\hat{\vec{P}}(x)-\frac{1}{c}\frac{\partial}{\partial t}\hat{\vec{A}}_{\text{radiation}}(x)\right)Z_{\alpha}e\hat{N}_{\alpha}(x)+\hat{\vec{\tau}}^{S}_{\alpha}(x)\\
&=\left(\hat{\vec{D}}(x)-\frac{1}{c}\frac{\partial}{\partial t}\hat{\vec{A}}_{\text{radiation}}(x)\right)Z_{\alpha}e\hat{N}_{\alpha}(x)-4\pi\hat{\vec{P}}(x)Z_{\alpha}e\hat{N}_{\alpha}(x)+\hat{\vec{\tau}}^{S}_{\alpha}(x)
\end{aligned} \tag{1.141}$$

Now we define (Tachibana 2002)

$$\hat{\overleftrightarrow{Z}}^{*}_{\alpha}(x)=Z_{\alpha}+\hat{\overleftrightarrow{Z}}_{\alpha\ \text{wind}}(x) \tag{1.142}$$

$$\hat{\overleftrightarrow{Z}}_{\alpha\ \text{wind}}(x)=\hat{\overleftrightarrow{Z}}_{\alpha\ \text{static wind}}(x)+\hat{\overleftrightarrow{Z}}_{\alpha\ \text{dynamic wind}}(x) \tag{1.143}$$

and we conclude the response tensor operators

$$\hat{\overleftrightarrow{Z}}_{\alpha\ \text{static wind}}(x) = -4\pi Z_\alpha \hat{\overleftrightarrow{\alpha}}(x) \tag{1.144}$$

$$\hat{\overleftrightarrow{Z}}_{\alpha\ \text{dynamic wind}}(x) e\left(\hat{\vec{D}}(x) - \frac{1}{c}\frac{\partial}{\partial t}\hat{\vec{A}}_{\text{radiation}}(x)\right)\hat{N}_\alpha(x) = \hat{\vec{\tau}}^S_\alpha(x) \tag{1.145}$$

It should be noted that the formulation for $\alpha = \mathrm{e}$ leads to electronic conduction. The usual textbook approach demonstrates the medium effect as the dissipative force against Lorentz force: see, e.g., Eq. (1.16) of the Ashcroft–Mermin textbook on solid-state physics (Ashcroft and Mermin 1976). In our present result, the dissipative force emerges from the tension density as the field theoretical force density compensating the Lorentz force density (see Fig. 1.30).

It should be noted that the response tensor $\overleftrightarrow{R}^{\Pi}(x)$, such as the electronic dielectric constant $\overleftrightarrow{\varepsilon}(\vec{r})$, the magnetic permeability $\overleftrightarrow{\mu}(\vec{r})$, the conductivity $\overleftrightarrow{\sigma}_{\text{int,ext}}(x)$, the index of refraction $\overleftrightarrow{n}(\vec{r})$, and the effective charge $\overleftrightarrow{Z}(\vec{r})$, is symmetry polarized in general inhomogeneous media. Mathematically, the response should be studied through the Jordan normal form of the symmetry-polarized response tensor operator $\hat{\overleftrightarrow{R}}^{\Pi}(x)$. More intuitively, the physical meaning of response may be revealed through the directional and rotational responses corresponding to the major elements of the symmetric $\hat{\overleftrightarrow{R}}^{S}(x)$ and antisymmetric $\hat{\overleftrightarrow{R}}^{A}(x)$ responses, respectively (see Fig. 1.31).

Let the total angular momentum may be conserved globally, Eq. (1.80), yet the particles may be locally deflected by inhomogeneity of the system. Actually, the

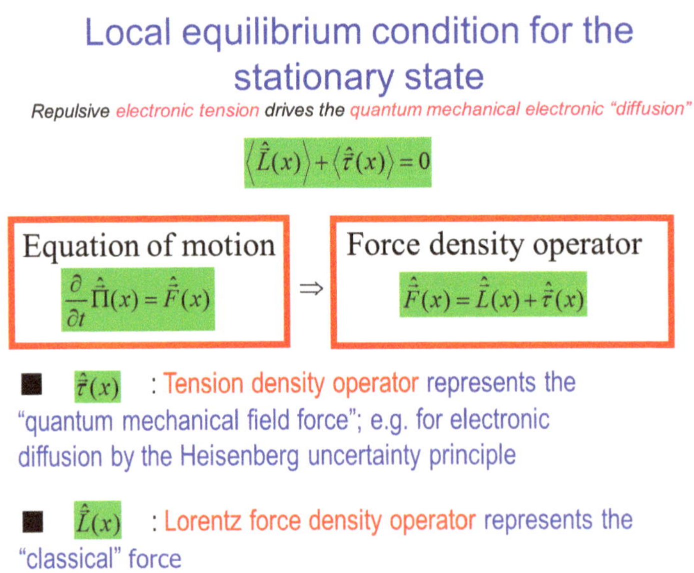

Fig. 1.30 Local equilibrium condition of the stationary state

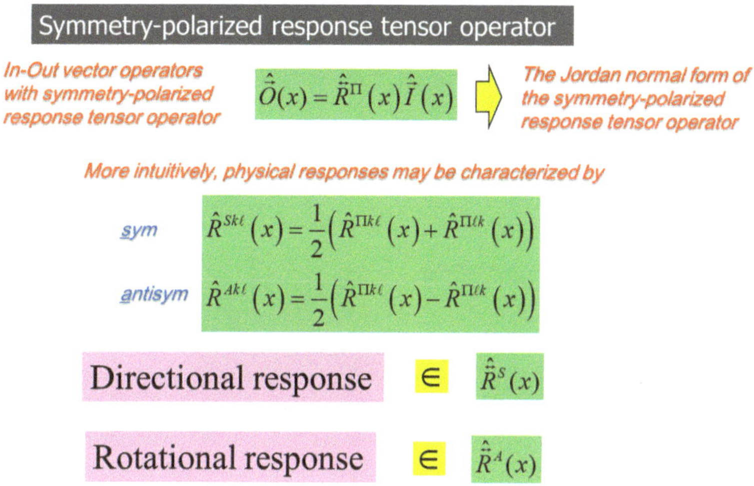

Fig. 1.31 Response tensor operator $\hat{\overleftrightarrow{R}}^{\Pi}(x)$

complex eigenvalue of the electronic dielectric constant $\overleftrightarrow{\varepsilon}(\vec{r})$ has been demonstrated numerically (Doi et al. 2006). The rotational response of electron toward applied electric field, namely, the electron deflection, should then be realized by the complex eigenvalue of $\overleftrightarrow{\varepsilon}(\vec{r})$. In general, the rotational deflected response of electron toward applied electromagnetic field should be realized by the complex eigenvalues of the electronic dielectric constant $\overleftrightarrow{\varepsilon}(\vec{r})$, magnetic permeability $\overleftrightarrow{\mu}(\vec{r})$, effective charge $\overleftrightarrow{Z}(\vec{r})$, and electric conductance $\overleftrightarrow{\sigma}(\vec{r})$ (Tachibana 2010).

1.4 Examples

1.4.1 Torque in Analytical Examples

1.4.1.1 Spin Torque in Free Space

We may first examine free particle satisfying the Dirac equation

$$(i\hbar\partial\!\!\!/ - mc)\psi(x) = 0 \tag{1.146}$$

where the generic mass m of the Dirac particle denotes m_e for electron. The stationary state solution with the third eigenvalue $\zeta = \pm\frac{1}{2}\hbar$ of spin $S^3 = \vec{S} \bullet \vec{e}_z$ using the unit vector $\vec{e}_z$ along the third axis is

$$\psi(x) = u(\vec{p}, \zeta)e^{-\frac{i}{\hbar}x\cdot p}, \quad (p\!\!\!/ - mc)u(\vec{p}, \zeta) = 0, \quad \zeta = \pm\frac{1}{2}\hbar \tag{1.147}$$

with

$$u\left(\vec{p},\frac{1}{2}\hbar\right)=\frac{1}{2\sqrt{p^0(p^0+mc)}}\begin{pmatrix}p^0+mc+p_z\\ p_+\\ p^0+mc-p_z\\ -p_+\end{pmatrix},\quad p_+=p_x+ip_y \tag{1.148}$$

$$u\left(\vec{p},-\frac{1}{2}\hbar\right)=\frac{1}{2\sqrt{p^0(p^0+mc)}}\begin{pmatrix}p_-\\ p^0+mc-p_z\\ -p_-\\ p^0+mc+p_z\end{pmatrix},\quad p_-=p_x-ip_y \tag{1.149}$$

In the rest frame attached to the Dirac particle, the charge density and the chiral spin density are

$$\begin{aligned}N_R\left(\vec{0},\pm\frac{1}{2}\hbar\right)&=\frac{1}{2}\\ N_L\left(\vec{0},\pm\frac{1}{2}\hbar\right)&=\frac{1}{2}\end{aligned} \tag{1.150}$$

$$\begin{aligned}\vec{\sigma}_R\left(\vec{0},\pm\frac{1}{2}\hbar\right)&=\pm\frac{1}{2}\vec{e}_z\\ \vec{\sigma}_L\left(\vec{0},\pm\frac{1}{2}\hbar\right)&=\pm\frac{1}{2}\vec{e}_z\end{aligned} \tag{1.151}$$

In the inertial frame attached to observer, we have instead

$$\begin{aligned}N_R\left(\vec{p},\pm\frac{1}{2}\hbar\right)&=\frac{1}{2p^0}(p^0\pm p_z)\\ N_L\left(\vec{p},\pm\frac{1}{2}\hbar\right)&=\frac{1}{2p^0}(p^0\mp p_z)\end{aligned} \tag{1.152}$$

$$\begin{aligned}\vec{\sigma}_R\left(\vec{p},\pm\frac{1}{2}\hbar\right)&=\pm\frac{mc}{2p^0}\vec{e}_z+\frac{1\pm\dfrac{p_z}{p^0+mc}}{2p^0}\vec{p}\\ \vec{\sigma}_L\left(\vec{p},\pm\frac{1}{2}\hbar\right)&=\pm\frac{mc}{2p^0}\vec{e}_z-\frac{1\mp\dfrac{p_z}{p^0+mc}}{2p^0}\vec{p}\end{aligned} \tag{1.153}$$

where the spin-orbit coupling appears in the chiral spin density, with polarization

$$\vec{s}\left(\vec{p},\pm\frac{1}{2}\hbar\right)=\frac{1}{2}\hbar\vec{\sigma}\left(\vec{p},\pm\frac{1}{2}\hbar\right)=\pm\frac{1}{2}\hbar\left(\frac{mc}{p^0}\vec{e}_z+\frac{p_z}{p^0(p^0+mc)}\vec{p}\right) \tag{1.154}$$

The spin torque does not of course work in this case, but if electron is accelerated by the external electromagnetic field, further spin-orbit coupling, the Thomas precession, and therefore the spin torque emerge to bring about the resultant further polarization.

The charge density, spin density, current, chiral current, and the zeta potential are then

$$j^{\mu}\left(\vec{p},\pm\frac{1}{2}\hbar\right)=cq\left(1,\frac{p_x}{p^0},\frac{p_y}{p^0},\frac{p_z}{p^0}\right) \tag{1.155}$$

$$j_5{}^{\mu}\left(\vec{p},\pm\frac{1}{2}\hbar\right)=\pm cq\left(\frac{p_z}{p^0},\frac{p_zp_x}{p^0(p^0+mc)},\frac{p_zp_y}{p^0(p^0+mc)},1+\frac{p_z{}^2}{p^0(p^0+mc)}\right) \tag{1.156}$$

$$\phi_5\left(\vec{p},\pm\frac{1}{2}\right)=\pm\frac{\hbar c^2}{2}\frac{p_z{}^z}{p^0} \tag{1.157}$$

where the generic charge q denotes q_e for electron. The torque and zeta force are calculated to be zero:

$$\vec{\zeta}=0,\quad \vec{t}=0 \tag{1.158}$$

and hence the sum

$$\frac{\partial}{\partial t}\vec{s}=\vec{t}+\vec{\zeta}=0 \tag{1.159}$$

which should be so since the state here is chosen stationary.

Now we have the null vorticity:

$$\mathrm{rot}\,\vec{s}=0 \tag{1.160}$$

The null vorticity does not contribute to the kinetic momentum.

1.4.1.2 Plane Wave Radiation Field

The Volkov solution of the Dirac particle under a plane-wave radiation field (see Fig. 1.32)

$$A^{\mu}=A^{\mu}(\phi),\quad \phi=k\cdot x=k^0ct-\vec{k}\bullet\vec{r},\quad \lim_{\phi\to\phi_{\mathrm{inf}}}A^{\mu}(\phi)=0 \tag{1.161}$$

is given as (Volkov 1935; Berestetskii et al. 1982)

$$\psi=\left(1+\frac{1}{2k\cdot pc}\frac{q}{c}\not{k}\not{A}\right)e^{\frac{i}{\hbar}S_0}u \tag{1.162}$$

The Dirac equation

$$\left(i\hbar \not{D}(x) - mc\right)\psi(x) = 0, \quad \not{D}(x) = \gamma^{\mu} D_{\mu}(x)$$

$$D_{\mu}(x) = \partial_{\mu} + i\frac{q}{\hbar c} A_{\mu}(x)$$

Plane wave gauge with boundary condition

$$A^{\mu} = A^{\mu}(\phi), \quad \phi = k \cdot x = k^{0}ct - \vec{k} \bullet \vec{r}, \quad \lim_{\phi \to \phi_{\text{inf}}} A^{\mu}(\phi) = 0$$

$$\lim_{\phi \to \phi_{\text{inf}}} \zeta = \pm\frac{1}{2}\hbar$$

Fig. 1.32 The Volkov solution of the Dirac particle under a plane-wave radiation field

$$S_0 = -x \cdot p - \int_{\phi_{\text{inf}}}^{\phi} \left(\frac{1}{k \cdot p} \frac{q}{c} p \cdot A - \frac{1}{2k \cdot p} \left(\frac{q}{c}\right)^2 A^2 \right) d\phi \tag{1.163}$$

$$(\not{p} - mc)u = 0, \quad \partial \cdot u = 0 \tag{1.164}$$

$$p^2 = (mc)^2 \tag{1.165}$$

Let the asymptotic free boundary condition with the third eigenvalue $\zeta = \pm\frac{1}{2}\hbar$ of spin $S^3 = \vec{S} \bullet \vec{e}_z$ be

$$\lim_{\phi \to \phi_{\text{inf}}} \zeta = \pm\frac{1}{2}\hbar \tag{1.166}$$

Then we have

$$j^{\mu}\left(\vec{p}, \pm\frac{1}{2}\hbar\right) = cq\frac{1}{p^0}\left(p^{\mu} - \frac{q}{c}A^{\mu} + k^{\mu}\left(\frac{1}{k \cdot p}\frac{q}{c}A \cdot p - \frac{1}{2k \cdot p}\left(\frac{q}{c}\right)^2 A^2\right)\right) \tag{1.167}$$

$$j_5{}^0\left(\vec{p}, \pm\frac{1}{2}\hbar\right) = \pm cq \begin{pmatrix} \dfrac{p_z}{p^0} \\ +\dfrac{1}{2k \cdot p}\dfrac{q}{c}\begin{pmatrix} -2A^0\left(k^0\dfrac{p_z}{p^0} - \dfrac{1}{p^0(p^0 + mc)}p_z\vec{k} \bullet \vec{p} - \dfrac{mc}{p^0}k_z\right) \\ +2k^0\left(A^0\dfrac{p_z}{p^0} - \dfrac{1}{p^0(p^0 + mc)}p_z\vec{A} \bullet \vec{p} - \dfrac{mc}{p^0}A_z\right) \end{pmatrix} \\ -\left(\dfrac{1}{2k \cdot p}\dfrac{q}{c}\right)^2 2A^2k^0\left(k^0\dfrac{p_z}{p^0} - \dfrac{1}{p^0(p^0 + mc)}p_z\vec{k} \bullet \vec{p} - \dfrac{mc}{p^0}k_z\right) \end{pmatrix} \tag{1.168}$$

$$
{j_5}^1\left(\vec{p},\pm\frac{1}{2}\hbar\right)=\pm cq\begin{pmatrix}\frac{1}{p^0(p^0+mc)}p_zp_x\\+\frac{1}{2k\cdot p}\frac{q}{c}\begin{pmatrix}-2A_x\left(k^0\frac{p_z}{p^0}-\frac{1}{p^0(p^0+mc)}p_z\vec{k}\bullet\vec{p}-\frac{mc}{p^0}k_z\right)\\+2k_x\left(A^0\frac{p_z}{p^0}-\frac{1}{p^0(p^0+mc)}p_z\vec{A}\bullet\vec{p}-\frac{mc}{p^0}A_z\right)\end{pmatrix}\\-\left(\frac{1}{2k\cdot p}\frac{q}{c}\right)^2 2A^2k_x\left(k^0\frac{p_z}{p^0}-\frac{1}{p^0(p^0+mc)}p_z\vec{k}\bullet\vec{p}-\frac{mc}{p^0}k_z\right)\end{pmatrix} \tag{1.169}
$$

$$
{j_5}^2\left(\vec{p},\pm\frac{1}{2}\hbar\right)=\pm cq\begin{pmatrix}\frac{1}{p^0(p^0+mc)}p_zp_y\\+\frac{1}{2k\cdot p}\frac{q}{c}\begin{pmatrix}-2A_y\left(k^0\frac{p_z}{p^0}-\frac{1}{p^0(p^0+mc)}p_z\vec{k}\bullet\vec{p}-\frac{mc}{p^0}k_z\right)\\+2k_y\left(A^0\frac{p_z}{p^0}-\frac{1}{p^0(p^0+mc)}p_z\vec{A}\bullet\vec{p}-\frac{mc}{p^0}A_z\right)\end{pmatrix}\\-\left(\frac{1}{2k\cdot p}\frac{q}{c}\right)^2 2A^2k_y\left(k^0\frac{p_z}{p^0}-\frac{1}{p^0(p^0+mc)}p_z\vec{k}\bullet\vec{p}-\frac{mc}{p^0}k_z\right)\end{pmatrix} \tag{1.170}
$$

$$
{j_5}^3\left(\vec{p},\pm\frac{1}{2}\hbar\right)=\pm cq\begin{pmatrix}\frac{mc}{p^0}+\frac{1}{p^0(p^0+mc)}{p_z}^2\\+\frac{1}{2k\cdot p}\frac{q}{c}\begin{pmatrix}-2A_z\left(k^0\frac{p_z}{p^0}-\frac{1}{p^0(p^0+mc)}p_z\vec{k}\bullet\vec{p}-\frac{mc}{p^0}k_z\right)\\+2k_z\left(A^0\frac{p_z}{p^0}-\frac{1}{p^0(p^0+mc)}p_z\vec{A}\bullet\vec{p}-\frac{mc}{p^0}A_z\right)\end{pmatrix}\\-\left(\frac{1}{2k\cdot p}\frac{q}{c}\right)^2 2A^2k_z\left(k^0\frac{p_z}{p^0}-\frac{1}{p^0(p^0+mc)}p_z\vec{k}\bullet\vec{p}-\frac{mc}{p^0}k_z\right)\end{pmatrix} \tag{1.171}
$$

Assume then for simplicity, first, radiation field propagates along the third axis associated with the electric field along the first axis and the magnetic field along the second axis

$$A^\mu=(0,A_x,0,0) \tag{1.172}$$

$$k^\mu=\left(k^0,0,0,k^0\right) \tag{1.173}$$

$$\vec{E}=-\frac{1}{c}\frac{\partial\vec{A}}{\partial t}=\left(E_x,E_y,E_z\right)=\left(-k^0\frac{dA_x}{d\phi},0,0\right) \tag{1.174}$$

$$\vec{B}=\mathrm{rot}\vec{A}=\left(B_x,B_y,B_z\right)=\left(0,-k^0\frac{dA_x}{d\phi},0\right) \tag{1.175}$$

Radiation field

$$A^\mu = (0, A_x, 0, 0),\quad k^\mu = (k^0, 0, 0, k^0)$$

$$\vec{E} = -\frac{1}{c}\frac{\partial \vec{A}}{\partial t} = (E_x, E_y, E_z) = \left(-k^0 \frac{dA_x}{d\phi}, 0, 0\right)$$

$$\vec{B} = \text{rot}\vec{A} = (B_x, B_y, B_z) = \left(0, -k^0 \frac{dA_x}{d\phi}, 0\right)$$

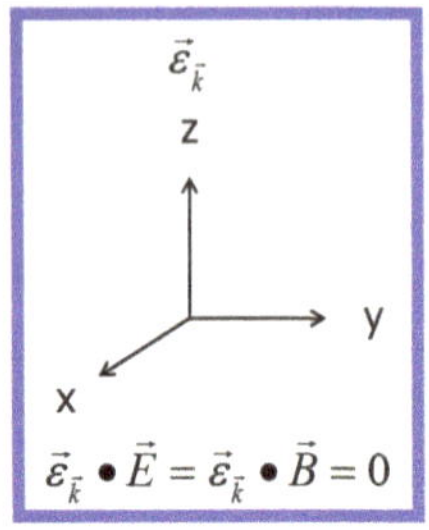

Momentum of the Dirac particle

$$p^\mu = (p^0, 0, 0, p_z)$$

Fig. 1.33 The plane-wave radiation field and momentum of the Dirac particle

and, second, the Dirac particle propagates along the third axis asymptotically (see Fig. 1.33)

$$p^\mu = (p^0, 0, 0, p_z) \tag{1.176}$$

It follows that the charge density, the spin density, and zeta potential are given as (see Fig. 1.34)

$$N = \frac{1}{cq} j^0 = 1 + \frac{1}{2p^0(p^0 - p_z)} \left(\frac{q}{c}\right)^2 (A_x)^2 \tag{1.177}$$

$$\vec{s} = \pm\frac{1}{2}\hbar\left(\frac{1}{p^0}\frac{q}{c}A_x, 0, 1 - (N - 1)\right) \tag{1.178}$$

$$\phi_5 = \pm\frac{\hbar c}{2}\left(\frac{p_z}{p^0} - (N - 1)\right) \tag{1.179}$$

The spin torque and zeta force are calculated to be

$$\vec{t} = (t_x, t_y, t_z) = \pm\frac{1}{2}\hbar\left(q\frac{k^0}{p^0}\frac{dA_x}{d\phi}, 0, 0\right) \tag{1.180}$$

$$\vec{\zeta} = (\zeta_x, \zeta_y, \zeta_z) = \pm\frac{1}{2}\hbar\left(0, 0, -\frac{ck^0}{p^0(p^0 - p_z)}\left(\frac{q}{c}\right)^2 A_x \frac{dA_x}{d\phi}\right) \tag{1.181}$$

Number density

$$N = \frac{1}{cq} j^0 = 1 + \frac{1}{2p^0\left(p^0 - p_z\right)} \left(\frac{q}{c}\right)^2 (A_x)^2$$

Spin density

$$\vec{s} = \pm \frac{1}{2} \hbar \left(\frac{1}{p^0} \frac{q}{c} A_x, 0, 1 - (N-1) \right)$$

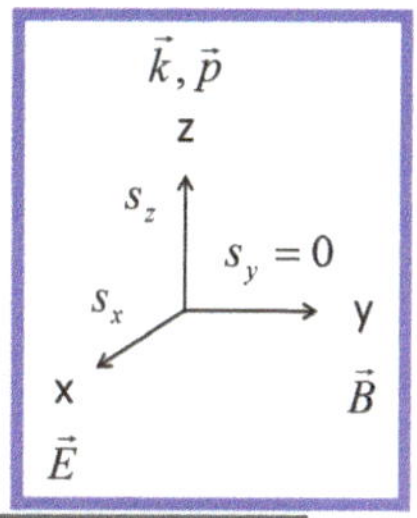

Spin torque and zeta force

$$\vec{t} = \left(\frac{\partial}{\partial t} s_x, 0, 0 \right), \quad \vec{\zeta} = -\mathrm{grad}\phi_5 = \left(0, 0, \frac{\partial}{\partial t} s_z \right)$$

Zeta potential

$$\phi_5 = \pm \frac{\hbar c}{2} \left(\frac{p_z}{p^0} - (N-1) \right)$$

Spin vorticity

$$\mathrm{rot}\vec{s} = \pm \frac{1}{2} \hbar \left(0, -\frac{k^0}{p^0} \frac{q}{c} \frac{dA_x}{d\phi}, 0 \right)$$

Fig. 1.34 Charge density, spin density, and zeta potential of the Dirac particle

Consequently, we have non-null spin dynamics, which should be so since the Volkov state is not stationary:

$$\frac{\partial}{\partial t} \vec{s} = \vec{t} + \vec{\zeta} \neq \vec{0} \tag{1.182}$$

The vorticity $\mathrm{rot}\,\vec{s}$ is

$$\mathrm{rot}\vec{s} = \pm \frac{1}{2} \hbar \left(0, -\frac{k^0}{p^0} \frac{q}{c} \frac{dA_x}{d\phi}, 0 \right) \tag{1.183}$$

Consequently, half the vorticity contributes to the kinetic momentum.

As a trivial limit of free electron in the stationary state, the torque and zeta force are calculated to be zero:

$$\vec{t} = \vec{0}, \quad \vec{\zeta} = \vec{0} \tag{1.184}$$

and hence the sum:

$$\frac{\partial}{\partial t} \vec{s} = \vec{t} + \vec{\zeta} = \vec{0} \tag{1.185}$$

which should be so since the state here is chosen stationary.

1.4.1.3 Static Uniform Magnetic Field

The Landau levels of the Dirac particle under a static uniform magnetic field along the third axis

$$A^{\mu}=\left(0,-\frac{1}{2}Hy,\frac{1}{2}Hx,0\right) \tag{1.186}$$

is given in a textbook (Greiner and Reinhardt 2009). Using the Landau eigenfunctions $R_{n,m_\ell,k_z,\sigma}(\rho)$ with $\rho=\sqrt{x^2+y^2}$, the torque and zeta force are calculated to be canceled with each other, which should be so since the state is stationary

$$\frac{\partial}{\partial t}\vec{s}=\vec{t}+\vec{\zeta}=\vec{0} \tag{1.187}$$

But the vector components are nonzero in this case:

$$\vec{\zeta}=-\mathrm{grad}\phi_5=\left(-\frac{\partial}{\partial x}\phi_5,-\frac{\partial}{\partial y}\phi_5,0\right) \tag{1.188}$$

with the zeta potential

$$\phi_5=\frac{\hbar c}{\dfrac{E_{n,m_\ell,k_z,\sigma}}{c}+mc}\frac{k_z\sigma}{(2\pi)^2}\left(R_{n,m_\ell,k_z,\sigma}(\rho)\right)^2 \tag{1.189}$$

where n and m_ℓ are the quantum numbers, k_z is the wave number along the third axis, and σ is the sign of the third eigenvalue $\zeta=\pm\frac{1}{2}\hbar$ of spin $S^3=\vec{S}\bullet\vec{e}_z$.

1.4.1.4 Spin Torque in Static Spherically Symmetric Scalar Potential

Here we examine static spherically symmetric scalar potential in hydrogen-like atom with the effective charge number Z_{eff}

$$A^{\mu}=\left(\frac{Z_{\mathrm{eff}}e}{r},0,0,0\right),\quad Z_{\mathrm{eff}}>0 \tag{1.190}$$

The stationary state solution in the Dirac representation is obtained in a textbook (Berestetskii et al. 1982) using the spherical coordinates r,θ,ϕ as

$$
\begin{aligned}
u_A &= f(r)\Omega_{j\ell m}, \quad \ell = j \pm \frac{1}{2} \\
u_B &= i^{1+\ell+\ell'} g(r)\Omega_{j\ell' m}, \quad \ell' = 2j - \ell \\
\Omega_{\ell\pm\frac{1}{2},\ell,m} &= \begin{pmatrix} \pm\sqrt{\dfrac{\ell+\frac{1}{2}\pm m}{2\ell+1}} {Y^L}_{\ell,m-\frac{1}{2}} \\ \sqrt{\dfrac{\ell+\frac{1}{2}\mp m}{2\ell+1}} {Y^L}_{\ell,m+\frac{1}{2}} \end{pmatrix} \\
{Y^L}_{\ell,m} &= i^\ell Y_{\ell,m} \\
&= (-)^m i^\ell \sqrt{\frac{(2\ell+1)(\ell-m)!}{4\pi(\ell+m)!}} P_\ell^m(\cos\theta) e^{im\phi}
\end{aligned}
\tag{1.191}
$$

with the energy eigenvalues

$$
\begin{aligned}
E_{n,j} &= mc^2 \frac{1}{\sqrt{1 + \dfrac{{Z_{eff}}^2\alpha^2}{\left(\sqrt{\kappa^2 - {Z_{eff}}^2\alpha^2} + n_r\right)^2}}}, \quad \alpha = \frac{e^2}{\hbar c} \\
\kappa &= \mp\left(j + \frac{1}{2}\right) = \begin{cases} -(\ell+1) \\ \ell \end{cases}, \quad j = \ell \pm \frac{1}{2} \\
n_r &= \begin{cases} 0, 1, 2, 3, \cdots, & \kappa < 0 \\ 1, 2, 3, \cdots, & \kappa > 0 \end{cases} = n - \left(j + \frac{1}{2}\right)
\end{aligned}
\tag{1.192}
$$

In the stationary state, the zeta potential is calculated to be null:

$$
\phi_5 = 0 \tag{1.193}
$$

neither the torque nor zeta force:

$$
\vec{t} = 0, \quad \vec{\zeta} = 0 \tag{1.194}
$$

Thus, as a matter of course, we have the stationary state of spin:

$$
\frac{\partial}{\partial t}\vec{s} = \vec{t} + \vec{\zeta} = 0 \tag{1.195}
$$

1.4.2 Torque in Molecules

1.4.2.1 Torque in Chiral Molecules

The spin torque, the zeta force, and the zeta potential, which are significant quantities to describe the local picture of spin dynamics of electron, are studied by using allene-type molecules, an achiral molecule C_3H_4 and a chiral molecule $C_3H_2Li_2$ (Fukuda et al. 2013). The two molecules have different distribution patterns of these quantities though their structures are similar to each other. It is also shown that the zeta potential distribution is almost independent of the electron density distribution (see Fig. 1.35).

Zeta potential is studied from the viewpoint of canonical orbitals (Fukuda et al. 2016a). Numerical example is C_6H_6 (see Fig. 1.36).

The local spin dynamics of electron is studied from the viewpoint of the electric dipole moment (EDM) of electron (Fukuda et al. 2016b). Numerical example is YbF (see Fig. 1.37).

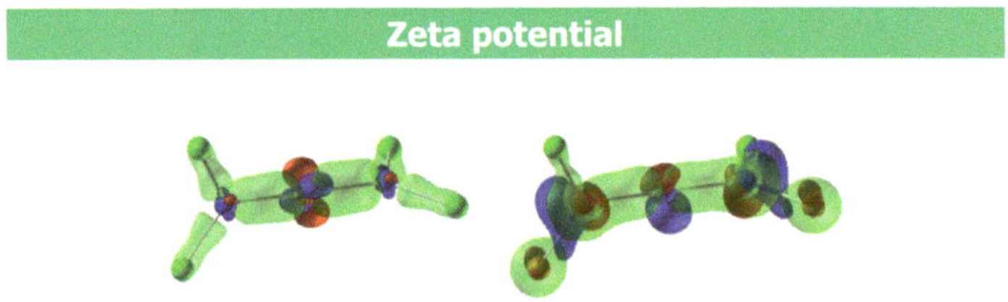

Fig. 1.35 Electron density and zeta potential in C_3H_4 and $C_3H_2Li_2$. *Blue* and *red* envelopes represent positive and negative zeta potential iso-surfaces, respectively. The threshold value of iso-surfaces of the zeta potential is taken as $\pm 7.5 \times 10^{-6}$ [a.u.]. *Green* envelopes represent electron density iso-surfaces. The threshold value of iso-surfaces of the electron density is taken as 0.25 [a.u.]

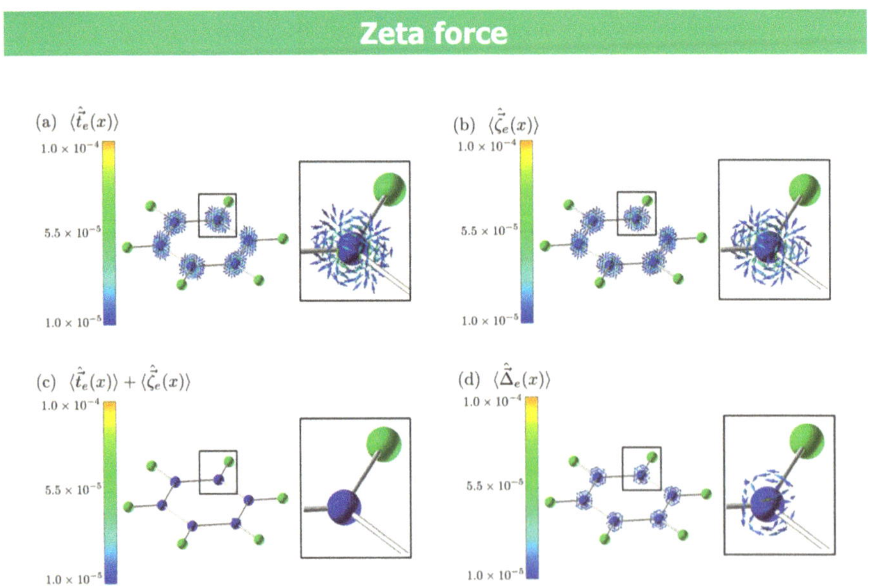

Fig. 1.36 The distribution of (**a**) the spin torque density, (**b**) the zeta force density, (**c**) the sum of them, and (**d**) the difference of large contributions in canonical orbitals in C_6H_6

Spin torque

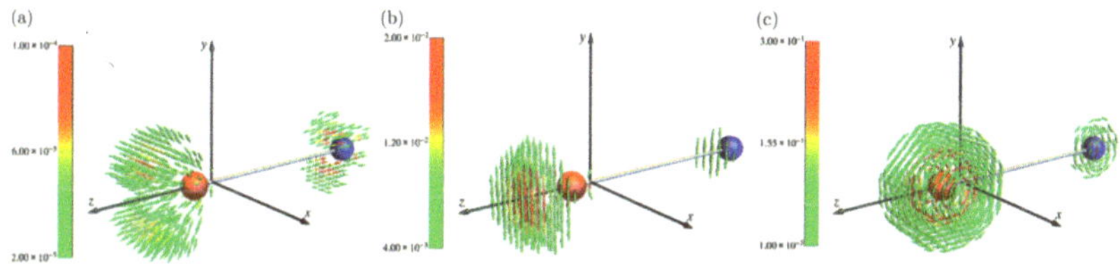

Fig. 1.37 Distributions of (**a**) the vector potential term of the spin torque density, (**b**) the electric term of the EDM torque density, and (**c**) the magnetic term of the EDM torque density in YbF. The *red* sphere represents the Yb nucleus, and the *blue* one represents the F nucleus. The *color* shows the value of the torque in atomic units

Spin vorticity

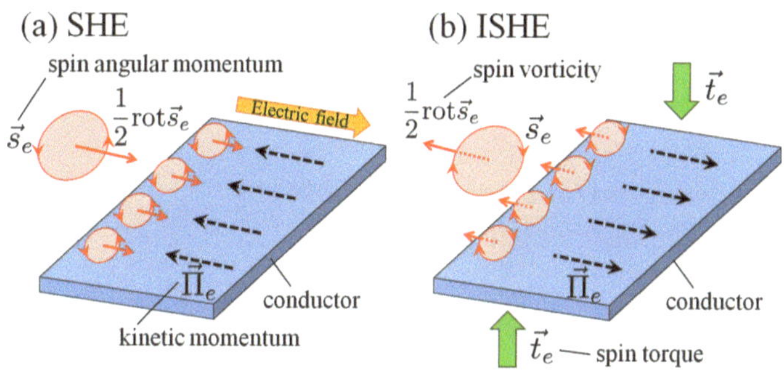

Fig. 1.38 Concept based on the quantum spin vorticity theory for (**a**) SHE and (**b**) ISHE

1.4.2.2 Spin Vorticity in Molecules

The spin vorticity of electron is studied from the viewpoint of the spin Hall effect (SHE) and the inverse spin Hall effect (ISHE) (Fukuda et al. 2016c). Idea here is the spin dynamics which may be realized in the bulk (see Fig. 1.38).

The realization of the bulk effect may be demanding. So numerical example is a straight carbon chain with bond length of 1.5 Å under a finite bias voltage of 0.1 V under an electronic temperature of 300 K (see Fig. 1.39).

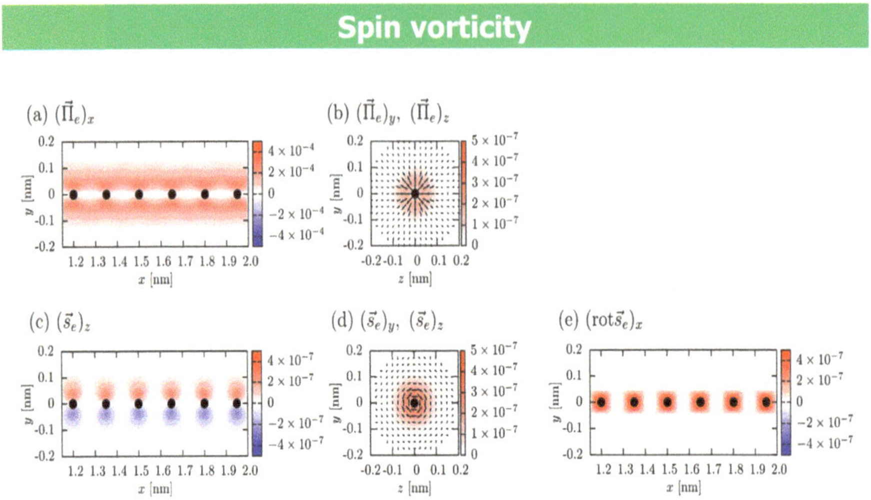

Fig. 1.39 (**a**) The distributions of the x component of the kinetic momentum density on the plane $z = 0$ [nm] and (**b**) y and z components on the plane $x = 1.65$ [nm]. (**c**) The distributions of the z component of the spin angular momentum density on the plane $z = 0$ [nm] and (**d**) y and z components on the plane $x = 1.65$ [nm]. (**e**) The distribution of the x component of the spin vorticity on the plane $z = 0$ [nm]. The y and z components of the spin vorticity on the plane $z = 0$ [nm] are negligibly small. In panels (**b**) and (**d**), the vectors consist of y and z components, and the color maps represent the norm of the vectors

1.4.3 Electromagnetic Properties of Matter in Magnetodielectric Media

It is confirmed numerically that the tension density defined in quantum field theory is the counter force to the Lorentz force density (Nozaki et al. 2016). Numerical example is benzenedithiol (see Fig. 1.40).

We use a nonequilibrium steady state model (Ikeda et al. 2013) as an example for system A embedded in an environmental background medium M (see Fig. 1.26). The response of electric current to electric field at a specific point in Si nanowire (see Fig. 1.41) does not have corresponding macroscopic physical quantity (Nozaki et al. 2016).

There are regions which show complicated response of electric current density to electric field, in particular, opposite and rotational ones (see Fig. 1.31). Local conductivities are considered to be available for the study of a negative differential resistance (NDR), which may be related to this opposite response (Ikeda et al. 2012). Numerical example is the Ge-substituted Si nanowire model (see Fig. 1.42).

Effective charge number of electromigration is studied for reliability problems of ultralarge-scale integration devices where extremely high current densities should be maintained through ultrathin film interconnects (Doi et al. 2003). Quantum mechanical wave-packet propagation of an Al atom has been examined in some models of thin Al lines which contain atomic defects, using the first-principle

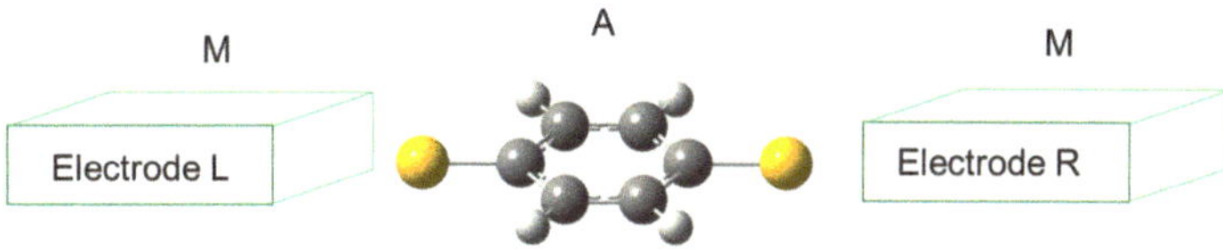

Fig. 1.40 Benzenedithiol A connected to external electrodes M

Fig. 1.41 Si nanowire models. *Light gray*, *green*, and *white* spheres correspond to Si, Ge, and H atoms, respectively

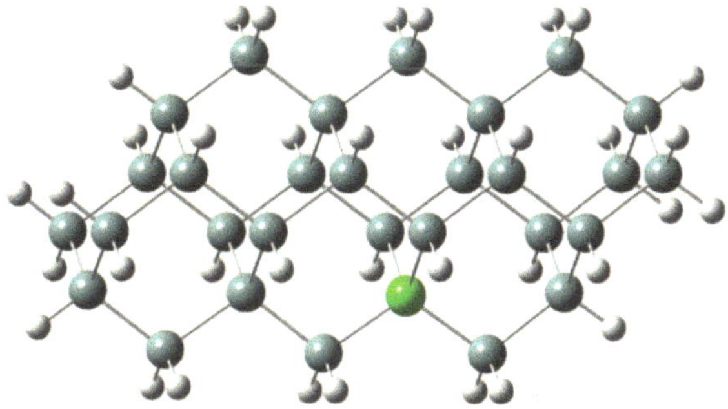

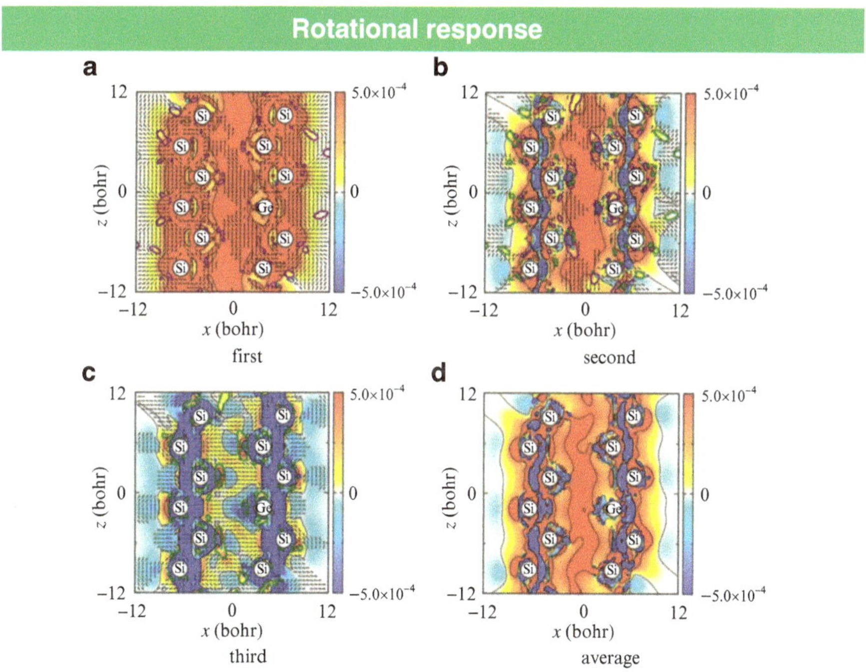

Fig. 1.42 Complex eigenvalues of (**a**) first, (**b**) second, (**c**) third, and (**d**) the average $\overleftrightarrow{\sigma}_{\mathrm{int}}(\vec{r})$ (a.u.) for the Ge-substituted model Si nanowire model (see Fig. 1.41)

electronic structure calculations under the periodic boundary condition (see Fig. 1.43).

The dynamic wind charge demonstrates significant figure at some characteristic point (see Fig. 1.44).

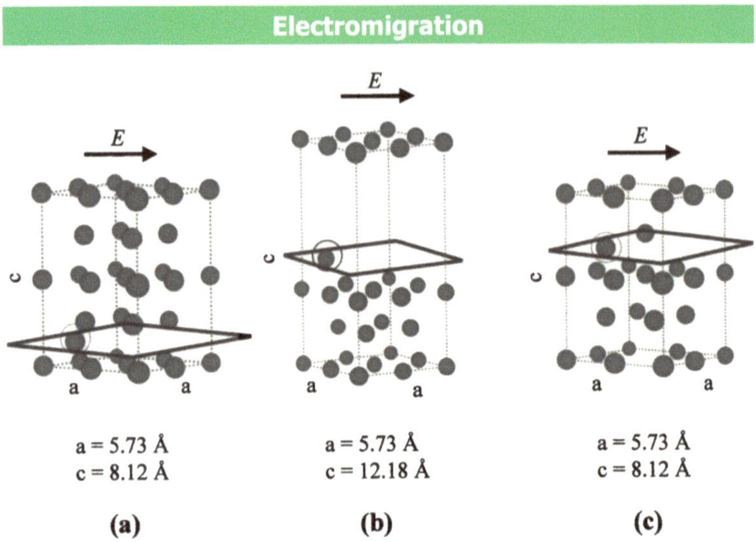

Fig. 1.43 Periodic models of electromigration in Al (100) surface for (**a**) bulk, (**b**) surface, and (**c**) grain boundary. *Arrows point* the direction of the external electric field, and squares inserted in the models indicate planes for maps in Fig. 1.33 on which the wave-packet ion core is put

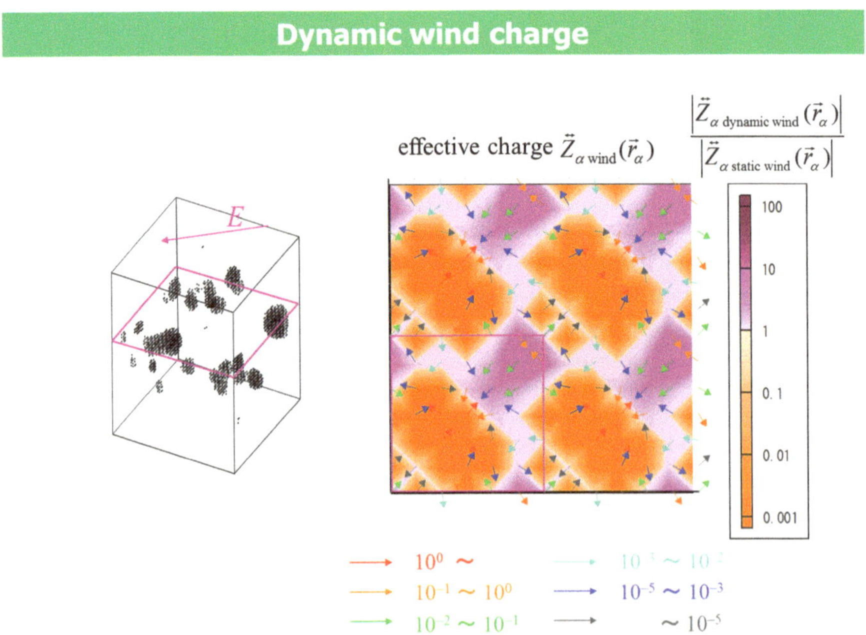

Fig. 1.44 Maps of the dynamic wind charge tensor density compared with the static one

1.5 Summary

Under external source of electromagnetic fields, charged particles can be accelerated by Lorentz force. Dissipative force can make the state of the charged particles stationary. Tension density of QED is formulated in such a way that it can compensate the Lorentz force density at any point of space-time. This formulation can give mechanical description of local equilibrium leading to the quantum mechanical stationary state.

The tension density is given by the divergence of stress tensor density. Electronic spin can be accelerated by torque density derived from the stress tensor density. The torque density can be compensated by a force density, called the zeta force density, which is another basic mechanism leading to the stationary state of the spinning motion of electron. It should be noted that the Pauli Hamiltonian gives equation of motion of electronic spin: see, e.g., Eq. (11.155) of the Jackson textbook on classical electrodynamics (Jackson 1998). The Bargmann–Michel–Telegdi (BMT) equation and Thomas precession are also the textbook matters. Our present result incorporates all of them in a closed form plus the field theoretical compensation mechanism leading to the stationary state of electronic spin. The external effect for chemical reaction systems is realized where a chemical reaction system A embedded in the environmental medium M is modeled as a parallel-plate capacitor filled with a dielectric. The vibronic interaction that goes beyond the adiabatic approximation has been incorporated as well as the electronic spin-dependent interaction.

References

N.W. Ashcroft, N.D. Mermin, *Solid State Physics* (Thomson Learning, New York, 1976) Chapter 1

V. Bargmann, E.P. Wigner, Proc. Natl. Acad. Sci. U. S. A. **34**, 211 (1948)

V.B. Berestetskii, E.M. Lifshitz, L.P. Pitaevskii, *Quantum Electrodynamics* (Butterworth-Heinemann, Oxford, 1982)

R.A. Beth, Phys. Rev. **48**, 471 (1935)

N.N. Bogoliubov, A.A. Logunov, L.T. Todorov, *Introduction to Axionatic Quantum Field Theory* (Benjamin, New York, 1975)

C. Bosvieux, J. Friedel, J. Phys. Chem. Solids **23**, 123 (1962)

R.P. Crease, *The Prism and the Pendulum: The Ten Most Beautiful Experiments in Science* (Random House, New York, 2006)

K. Doi, K. Iguchi, K. Nakamura, A. Tachibana, Phys. Rev. B **67**, 115124(14) (2003)

K. Doi, K. Nakamura, A. Tachibana, *2006 International Workshop on Nano CMOS* (IEEE 2006) pp. 209–235

P. Fara, Phil. Trans. R. Soc. A **373**, 20140213 (2015)

R.P. Feynman, R.B. Leighton, M. Sands, Quantum mechanics, in *The Feynman Lectures on Physics, Vol. **III*** (Addison-Wesley, Reading, 1972), Chapter 1, 1–1

M. Fukuda, M. Senami, A. Tachibana, in *Advances in Quantum Methods and Applications in Chemistry, Physics, and Biology Progress in Theoretical Chemistry and Physics*, ed by

M. Hotokka, E. J. Brändas, J. Maruani, G. D. Barrio. (Springer, New York, 2013) Chap.7, pp.131–139
M. Fukuda, K. Soga, M. Senami, A. Tachibana, Int. J. Quant. Chem. **116**, 920 (2016a)
M. Fukuda, K. Soga, M. Senami, A. Tachibana, Phys. Rev. A **93**, 012518(10) (2016b)
M. Fukuda, K. Ichikawa, M. Senami, A. Tachibana, AIP Adv. **6**, 025108(8) (2016c)
W. Greiner, J. Reinhardt, *Quantum Electrodynamics* (Springer, Berlin, 2009)
R. Haag, *Local Quantum Physics – Fields, Particled, Algebras*, 2nd edn. (Springer, New York, 1992)
W. Heitler, *The Quantum Theory of Radiation*, 3rd edn. (Oxford University, Oxford, 1954)
Y. Ikeda, M. Senami, A. Tachibana, AIP Adv. **2**, 042168(16) (2012)
Y. Ikeda, M. Senami, A. Tachibana, Trans. Mat. Res. Soc. Japan **38**, 397 (2013)
C. Itzykson, J.B. Zuber, *Quantum Field Theory* (McGraw-Hill, New York, 1980)
J.D. Jackson, *Classical Electrodynamics* (Wiley, New York, 1998)
L.D. Landau, E.M. Lifshitz, *The Classical Theory of Fields*, 6th edn. (Nauka, Moscow, 1973), translated into English in *Course of Theoretical Physics Series, Vol. 2*, 4th edn. (Pergamon, Oxford, 1975)
L.D. Landau, E.M. Lifshitz, in *The Classical Theory of Fields*. Course of Theoretical Physics Series, vol. **2**, 4th edn. (Pergamon, Oxford, 1975)
U. Leonhardt, Nature **444**, 823 (2006)
A. Lodder, J.P. Dekker, in *Stress Induced Phenomena in Metallization*, ed by H. Okabayashi, S. Shingubara, P. S. Ho. (American Institute of Physics, New York, 1998), pp. 315–328
C.W. Misner, K.S. Thorne, J.A. Wheeler, *Gravitation* (W. H. Freeman, San Fransisco, 1973)
N. Nakanishi, I. Ojima, *Covariant Operator Formalism of Gauge Theories and Quantum Gravity* (World Scientific, Singapore, 1990)
H. Nozaki, M. Senami, K. Ichikawa, A. Tachibana, Jpn. J. Apl. Phys. **55**, 08PE01(8) (2016)
W. Pauli, *Handbuch der Physik*, Band **XXIV**, Teil 1 (Springer, Berlin, 1933), pp. 83–272; reprinted in *Handbuch der Physik* (Springer, Berlin, 1958), Vol. **5**, Part1; translated into English in *General Principles of Quantum Mechanics* (Springer, Berlin, 1980)
M.E. Peskin, D.V. Schroeder, *An Introduction to Quantum Field Theory* (Westview, Boulder, 1995)
W. Poyser, *Magnetism and Electricity: A Manual for Students in Advanced Classes* (Longmans, Green, & Co., New York, 1892), p. 285
T. Rothman, *Everything's Relative and Other Fables in Science and Technology* (Wiley, New York, 2003)
L.H. Ryder, *Quantum Field Theory* (Cambridge University, Cambridge, 1985)
J.J. Sakurai, *Advnced Quantum Mechanics* (Addison-Wesley, New York, 1967)
E. Schödinger, Ann. Phys. (Leipzig) **82**, 265 (1927)
A. Tachibana, J. Chem. Phys. **115**, 3497–3518 (2001)
A. Tachibana, in *Stress Induced Phenomena in Metallization*, ed by S. P. Baker. (American Institute of Physics, New York, 2002), pp. 105–116
A. Tachibana, in *Fundamental Perspectives in Quantum Chemistry: A Tribute to the Memory of Per-Olov Löwdin*, ed by E. Brändas, E. Kryachko, vol 2 (Kluwer, Dordrecht, 2003), pp. 211–239
A. Tachibana, Int. J. Quant. Chem. **100**, 981 (2004)
A. Tachibana, J. Mol. Struct. (THEOCHEM) **943**, 138 (2010)
A. Tachibana, J. Math. Chem. **50**, 669 (2012)
A. Tachibana, in *Concepts and Methods in Modern Theoretical Chemistry: Electronic Structure and Reactivity (Atoms, Molecules, and Clusters)*, ed by S. K. Ghosh, P. K. Chattaraj. (Taylor & Francis/CRC Press, New York, 2013) , Chap.12, pp. 235–251
A. Tachibana, J.Comput.Chem.Jpn. **13**, 18 (2014)
A. Tachibana, J. Math. Chem. **53**, 1943 (2015)
A. Tachibana, J. Math. Chem. **54**, 661 (2016)
V.H. Tetrode, Z. Phys. **49**, 858 (1928)

D.M. Volkov, Z. Phys. **94**, 250 (1935)
S. Weinberg, *Gravitation and Cosmology: Principles and Applications of the General Theory of Relativity* (Wiley, New York, 1972)
S. Weinberg, *The Quantum Theory of Fields I-III* (Cambridge University, Cambridge, 1995)
E.P. Wigner, Ann. Math. **40**, 149 (1939)

Chapter 2
Energy-Momentum Tensor of QED

Abstract In Sect. 1.2.7, Chap. 1, it is found that the electron spin torque is counterbalanced by the chiral electron density. In Sect. 1.2.8, Chap. 1, it is found that the spin vorticity of electron contributes to the kinetic momentum of electron, which raises a simple but "odd" question: what is momentum of electron spin? In this Chapter, we shall show that the origin of both the chiral nature and the kinetic nature is manifest in the principle of equivalence in general relativity.

Keywords Chirality • Primary Rigged QED theory • Principle of equivalence • Rigged QED theory • Stress tensor • Spin torque • Spin vorticity • SUGRA • SUSY • Tension • Zeta force

2.1 Energy-Momentum Tensor

Light bends in order to advance the space-time that has been distorted by heavy mass objects (see Fig. 2.1). This is called the gravitational lens, a phenomenon which is predicted by the general theory of relativity. It is one of the phenomena that space-time has proven the curvature (Weinberg 1972; Hayashi and Shirafuji 1979; Nakanishi 2004). Dynamics of electrons as the vorticity contribution to the momentum is derived from the geometric principle associated with the tetrad field with torsion (Tachibana 2012). *The action must be generally covariant, with all fields treated as scalars, except for the tetrad field itself.* The Weitzenböck space-time is the key to warrant the tetrad field for the description of the Dirac spinor.

2.1.1 Principle of Equivalence

The most general setup of space-time for the Dirac spinor field in QED with the principle of equivalence (see Fig. 2.2) is the Riemann–Cartan space-time (see Fig. 2.3).

Torsion and curvature are the characteristics of the space-time geometry (see Fig. 2.4). The tetrad field is associated with the Dirac spinor field (Hehl et al. 1976)

A. Tachibana, *New Aspects of Quantum Electrodynamics*,
DOI 10.1007/978-981-10-3132-8_2

Fig. 2.1 X-ray selected sample of massive lensing clusters (Reproduced from Lopes 2011)

Fig. 2.2 Space-time structure

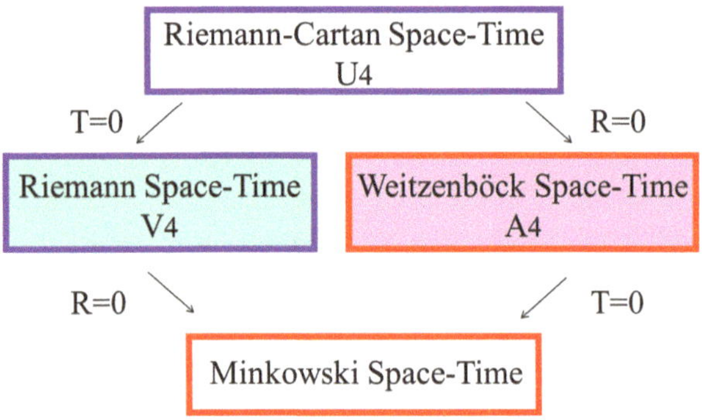

The curvature-free but non-null torsion Weitzenböck space-time is indispensable for the absolute parallelism of the tetrad field (see Fig. 2.5).

The Dirac spinor field is a coordinate scalar and a Lorentz spinor for the torsion-free Riemann space-time (Weinberg 1972). Supersymmetry (SUSY) is the nontrivial extension of the Poincaré algebra (Haag et al. 1975). The gauge boson of the localized SUSY is spin-2 graviton, where the theory of supergravity (SUGRA) emerges (Weinberg 1995). The quantum electron spin vorticity principle is the consequence of the principle of equivalence both in QED equipped with semiclassical Einstein–Hilbert action (Tachibana 2012) and simple SUGRA (Tachibana 2014).

Fig. 2.3 The Riemann–Cartan space-time characteristics

The Riemann-Cartan Space-Time U4

Metric covariant derivative

$$g_{\mu\nu;\lambda} = \partial_\lambda g_{\mu\nu} - \Gamma^{\rho}_{\mu\lambda} g_{\rho\nu} - \Gamma^{\rho}_{\nu\lambda} g_{\mu\rho}$$

Connection

$$\Gamma^{\lambda}_{\mu\nu} = \left\{ {\lambda \atop \mu\nu} \right\} + K^{\lambda}_{\bullet\;\mu\nu}$$

The Levi-Civita connection

$$\left\{ {\lambda \atop \mu\nu} \right\} = \frac{1}{2} g^{\lambda\rho} \left(\partial_\mu g_{\nu\rho} + \partial_\nu g_{\mu\rho} - \partial_\rho g_{\mu\nu} \right)$$

Contorsion tensor

$$K^{\lambda}_{\bullet\;\mu\nu} = \frac{1}{2} \left(T^{\lambda}_{\bullet\;\mu\nu} - T^{\bullet\lambda}_{\mu\bullet\nu} - T^{\bullet\lambda}_{\nu\bullet\mu} \right)$$

Fig. 2.4 Torsion and curvature of space-time geometry

Torsion and curvature

Torsion tensor

$$T^{\lambda}_{\bullet\;\mu\nu}(\Gamma) = \Gamma^{\lambda}_{\mu\nu} - \Gamma^{\lambda}_{\nu\mu}$$

Curvature tensor

$$R^{\rho}{}_{\sigma\mu\nu}(\Gamma) = \partial_\mu \Gamma^{\rho}_{\sigma\nu} - \partial_\nu \Gamma^{\rho}_{\sigma\mu} + \Gamma^{\rho}_{\lambda\mu} \Gamma^{\lambda}_{\sigma\nu} - \Gamma^{\rho}_{\lambda\nu} \Gamma^{\lambda}_{\sigma\mu}$$

The Ricci tensor

$$R_{\mu\nu}(\Gamma) = R^{\rho}{}_{\mu\nu\rho}(\Gamma)$$

Curvature

$$R(\Gamma) = R^{\mu}{}_{\mu}(\Gamma)$$

Fig. 2.5 The Weitzenböck space-time characteristics

The Weitzenböck Space-Time A4

Absolute parallerism of vierbein (tetrad) field

$$\partial_\nu e^{\lambda}_{a} + \Gamma^{*\lambda}_{\mu\nu} e^{\mu}_{a} = 0$$

$$g^{\mu\nu} = \eta^{ab} e^{\mu}_{a} e^{\nu}_{b}, \quad g^{\mu\nu} e^{a}_{\mu} e^{b}_{\nu} = \eta^{ab} = diag(1,-1,-1,-1)$$

Connection

$$\Gamma^{*\lambda}_{\mu\nu} = e^{\lambda}_{c} \partial_\nu e^{c}_{\mu}$$

Curvature tensor null

$$R^{\rho}{}_{\mu\nu\sigma}(\Gamma^*) = 0$$

Tortion tensor non-null

$$T^{\lambda}_{\bullet\;\mu\nu}(\Gamma^*) = e^{\lambda}_{b} \left(\partial_\nu e^{b}_{\mu} - \partial_\mu e^{b}_{\nu} \right)$$

2.1.1.1 The Einstein Tensor

To seek for the variation principle of the equation of motion on the background-curved space-time, the semiclassical Einstein–Hilbert action integral has been used under the symmetry of the general coordinate transformation of gravity

$$\delta I = 0, \quad I = \frac{c}{2\kappa}\int R\sqrt{-g}d^4x + \frac{1}{c}\int L\sqrt{-g}d^4x, \quad \kappa = \frac{8\pi G}{c^2} \tag{2.1}$$

where R is the Ricci scalar, G is the universal gravitational constant, and L is the Lagrangian density of QED including the interaction with gravity. The gravitational action I_G is added to the system action I_S and made stationary

$$\delta I = 0, \quad I = I_G + I_S \tag{2.2}$$

under the variation $\delta g^{\mu\nu}$ of the metric tensor $g^{\mu\nu}$

$$I_G = \frac{c}{2\kappa}\int R\sqrt{-g}d^4x, \quad \delta I_G = \frac{c}{2\kappa}\int \left(R_{\mu\nu} - \frac{1}{2}g_{\mu\nu}R\right)\delta g^{\mu\nu}\sqrt{-g}d^4x \tag{2.3}$$

$$I_S = \frac{1}{c}\int L\sqrt{-g}d^4x, \quad \delta I_S = \frac{1}{2c}\int T_{\mu\nu}\delta g^{\mu\nu}\sqrt{-g}d^4x \tag{2.4}$$

The Einstein equation is then derived (see Fig. 2.6)

$$G_{\mu\nu}(x) = Y_{\mu\nu}(x) \tag{2.5}$$

with the definition

$$G_{\mu\nu}(x) = \frac{1}{\sqrt{-g(x)}}\frac{\delta}{\delta g^{\mu\nu}(x)}\frac{2\kappa}{c}I_G = R_{\mu\nu}(x) - \frac{1}{2}g_{\mu\nu}(x) \tag{2.6}$$

$$Y_{\mu\nu}(x) = -\frac{1}{\sqrt{-g(x)}}\frac{\delta}{\delta g^{\mu\nu}(x)}\frac{2\kappa}{c}I_S = -\frac{\kappa}{c^2}T_{\mu\nu}(x) \tag{2.7}$$

Since the Einstein tensor $G_{\mu\nu}(x)$ is symmetric, so is the energy-momentum tensor $T_{\mu\nu}(x)$ (see Fig. 2.7).

$$G_{\mu\nu}(x) = G_{\nu\mu}(x); \text{ symmetric} \tag{2.8}$$

$$T_{\mu\nu}(x) = T_{\nu\mu}(x); \text{ symmetric} \tag{2.9}$$

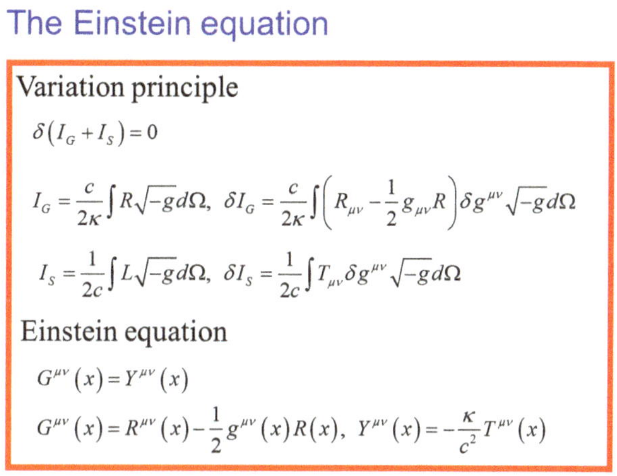

Fig. 2.6 The semiclassical Einstein–Hilbert action principle

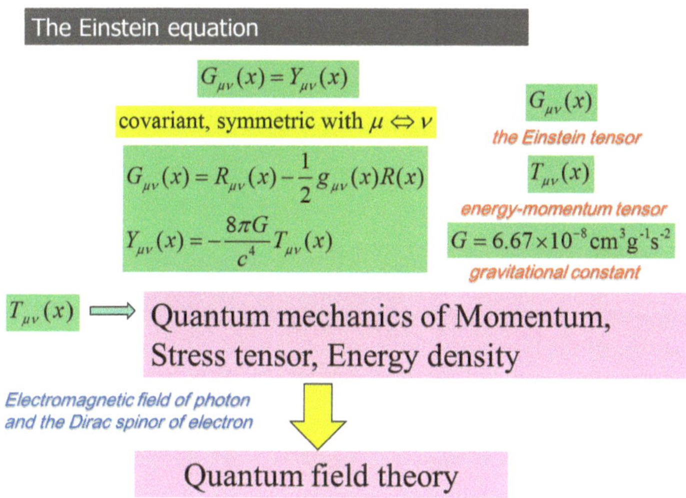

Fig. 2.7 The semiclassical Einstein–Hilbert field theory

2.1.1.2 Tetrad Formalism

Using *the tetrad formalism equipped with the principle of equivalence*, the metric tensor in any general noninertial coordinate system is given as

$$g_{\mu\nu}(x) = e^a{}_\mu(x)e^b{}_\nu(x)\eta_{ab} \tag{2.10}$$

where $e^a{}_\mu(x)$ denotes the tetrad field and the Latin letters a, b, c, and so on run from 0 to 3. The tetrad field $e^a{}_\mu(x)$ is a coordinate vector and a Lorentz vector for the Lorentz transformation $x \to x'$ associated with the vector representation $\Lambda^a{}_b(x)$

$$e^a{}_\mu(x) \to e'^a{}_\mu(x') = \frac{\partial x^\nu}{\partial x'^\mu} e^a{}_\nu(x) \tag{2.11}$$

$$e^a{}_\mu(x) \to e'^a{}_\mu(x) = \Lambda^a{}_b(x) e^b{}_\mu(x) \tag{2.12}$$

and is parallely transported

$$\partial_\nu e_a{}^\lambda + \left\{{}_\kappa{}^\lambda{}_\nu\right\} e_a{}^\kappa - \gamma_a{}^b{}_\nu e_b{}^\lambda = 0 \tag{2.13}$$

We have used the Levi–Civita affine connection

$$\left\{{}_\mu{}^\lambda{}_\nu\right\} = \frac{1}{2} g^{\lambda\rho} \left(\partial_\mu g_{\nu\rho} + \partial_\nu g_{\mu\rho} - \partial_\rho g_{\mu\nu} \right) = \left\{{}_\nu{}^\lambda{}_\mu\right\} \tag{2.14}$$

and spin connection

$$\gamma_a{}^b{}_\mu = e_{a\nu;\mu} \eta^{bc} e_c{}^\nu \tag{2.15}$$

where the covariant derivative is defined as

$$e_a{}^\lambda{}_{;\nu} = e_a{}^\lambda{}_{,\nu} + \left\{{}_\kappa{}^\lambda{}_\nu\right\} e_a{}^\kappa \tag{2.16}$$

$$e_{a\lambda;\nu} = e_{a\lambda,\nu} - \left\{{}_\lambda{}^\kappa{}_\nu\right\} e_{a\kappa} \tag{2.17}$$

with the usual partial derivative denoted as

$$f_{,\mu} = \partial_\mu f \tag{2.18}$$

In the tetrad formalism, the absolute parallelism of the tetrad field $e^a{}_\mu(x)$ is found to be

$$D^*{}_\nu e_a{}^\lambda = \partial_\nu e_a{}^\lambda + \Gamma^*{}_\mu{}^\lambda{}_\nu e_a{}^\mu = 0 \tag{2.19}$$

and the connection

$$\Gamma^*{}_\mu{}^\lambda{}_\nu = \left\{{}_\mu{}^\lambda{}_\nu\right\} - e^a{}_\mu \gamma_a{}^b{}_\nu e_b{}^\lambda \tag{2.20}$$

is used to define the torsion tensor

$$T^*{}^\lambda_{\cdot}\mu\nu = \Gamma^*{}_\mu{}^\lambda{}_\nu - \Gamma^*{}_\nu{}^\lambda{}_\mu \tag{2.21}$$

and contorsion tensor

$$K^{*}{}_{\lambda\mu\nu} = \frac{1}{2}\left(T^{*}{}_{\lambda\mu\nu} - T^{*}{}_{\mu\lambda\nu} - T^{*}{}_{\nu\lambda\mu}\right) \tag{2.22}$$

The Dirac spinor field is a coordinate scalar and a Lorentz spinor

$$\psi_\alpha(x) \to \psi'_\alpha(x') = \psi_\alpha(x) \tag{2.23}$$

$$\psi_\alpha(x) \to \psi'_\alpha(x) = D_{\alpha\beta}(\Lambda(x))\psi_\beta(x) \tag{2.24}$$

Also, what is important, the covariant derivative $D_\mu(g)$, is not only a coordinate scalar but also a Lorentz vector, as shown in Eqs. (12.5.15–12.5.17) and (12.5.24) of Weinberg (1972):

$$D_\mu(g) = \partial_\mu + \Gamma_\mu \tag{2.25}$$

$$\Gamma_\mu(x) \to \Gamma_\mu{}'(x) = D(\Lambda(x))\Gamma_\mu D^{-1}(\Lambda(x)) - \left(\partial_\mu D(\Lambda(x))\right)D^{-1}(\Lambda(x)) \tag{2.26}$$

The Lagrangian density for the QED system under external gravity is then given as

$$L = L_{\mathrm{EM}} + L_{\mathrm{Dirac}} \tag{2.27}$$

with the definition

$$L_{\mathrm{EM}} = -\frac{1}{16\pi}F_{\mu\nu}F^{\mu\nu} = -\frac{1}{16\pi}F_{\mu\nu}F_{\rho\sigma}g^{\mu\rho}g^{\nu\sigma}, \;\; F_{\mu\nu} = \partial_\mu A_\nu - \partial_\nu A_\mu \tag{2.28}$$

$$L_{\mathrm{Dirac}} = \frac{1}{2}c\overline{\psi}\left(i\hbar\gamma^a e_a{}^\mu D_\mu(g) - mc\right)\psi + h.c. \tag{2.29}$$

The gravitational covariant derivative$D_\mu(g)$ is concretely written as (see Fig. 2.8)

$$\begin{aligned} D_\mu(g) &= \partial_\mu + i\frac{1}{2\hbar}\gamma_{ab\mu}J^{ab} + i\frac{q}{\hbar c}A_\mu \\ &= D_\mu + i\frac{1}{2\hbar}\gamma_{ab\mu}J^{ab} \end{aligned} \tag{2.30}$$

where the spin angular momentum J^{ab}

$$J^{ab} = \frac{i\hbar}{4}\left[\gamma^a, \gamma^b\right] \tag{2.31}$$

is added to D_μ through the coupling with spin connection $\gamma_{ab\mu}$ given in Eq. (2.15). *The emergence of the spin connection is manifest as the consequence of the principle of equivalence in general relativity.*

It should be noted here that after some manipulation, we can rewrite Eq. (2.29) in a very significant form as follows:

Fig. 2.8 Gravity covariant derivative for the Dirac spinor field

The coordinate-scalar Lorentz-vector derivative

$$D_a = e_a^\mu \left(\partial_\mu + i\frac{\hbar}{2}\sigma_{bc}\omega_\mu^{bc} \right)$$

The spin angular momentum

$$J^{ab} = \frac{i\hbar}{4}\left[\gamma^a, \gamma^b\right]$$

The Ricci rotation coefficient

$$\omega_\mu^{ab} = \gamma_{cde}\eta^{cb}\eta^{da}e_\mu^e, \ \gamma_{abc} = e_{a\mu;\nu}e_b^\mu e_c^\nu$$

$$\begin{aligned} L_{\text{Dirac}} &= \frac{1}{2}c\overline{\psi}\left(i\hbar\gamma^a e_a{}^\mu D_\mu(g) - mc\right)\psi + h.c. \\ &= \frac{1}{2}c\overline{\psi}\left(i\hbar\gamma^a e_a{}^\mu \partial_\mu - mc\right)\psi + h.c. - \frac{3\hbar}{4q}a_\mu j_5{}^\mu - \frac{1}{c}A_\mu j^\mu \end{aligned} \tag{2.32}$$

Namely, which is hidden in Eq. (2.29), but in this Eq. (2.32), minimal couplings are manifestly shown; those not only of current $j^\mu(x)$ with photon vector potential $A^\mu(x)$ but also of chiral current $j_5^\mu(x)$ with spin coupling vector $a^\mu(x)$ defined as

$$a^\mu = \frac{1}{6}\varepsilon^{\mu\nu\rho\sigma}(g)T^*{}_{\nu\rho\sigma} \tag{2.33}$$

where $T^*{}_{\nu\rho\sigma}$ is the torsion tensor given in Eq. (2.21), and we have used the Levi–Civita tensor

$$\varepsilon^{\mu\nu\rho\sigma}(g) = \frac{1}{\sqrt{-g}}\varepsilon^{\mu\nu\rho\sigma}, \quad \varepsilon^{0123} = 1 \tag{2.34}$$

$$\varepsilon_{\mu\nu\rho\sigma}(g) = \sqrt{-g}\varepsilon_{\mu\nu\rho\sigma}, \quad \varepsilon_{0123} = -1 \tag{2.35}$$

Using the Lagrangian density given in Eq. (2.27), the variation principle with respect to the spinor field

$$\frac{\delta}{\delta\overline{\psi}}I_S = 0 \tag{2.36}$$

leads to the Dirac equation of the Dirac particle

$$\left(i\hbar\gamma^a e_a{}^\mu D_\mu(g) - mc\right)\psi = 0 \tag{2.37}$$

and similarly the Maxwell equation of photon

$$F^{\nu\mu}{}_{;\nu} = \frac{4\pi}{c} j^{\mu} \tag{2.38}$$

with the continuity equation of current

$$\partial_{\mu} j^{\mu} = 0 \tag{2.39}$$

Second, the variation principle with respect to the tetrad field leads to the symmetric energy-momentum tensor $T_{\mu\nu}$ and the conservation law as follows:

$$\delta I_S = \delta \frac{1}{c} \int L\sqrt{-g} d^4 x = \frac{1}{c} \int T^{\cdot a}_{\mu} \delta e_a{}^{\mu} \sqrt{-g} d^4 x \tag{2.40}$$

$$T_{\mu\nu}\sqrt{-g} = \eta_{ab} e^b{}_{\nu} \frac{\partial}{\partial e_a{}^{\mu}} L\sqrt{-g} \tag{2.41}$$

The symmetric energy-momentum tensor

$$T_{\mu\nu} = -\varepsilon^{\Pi}{}_{\mu\nu} - \tau^{\Pi}{}_{\mu\nu}(g) - \frac{1}{4\pi} g^{\rho\sigma} F_{\mu\rho} F_{\nu\sigma} - g_{\mu\nu}(L_{\mathrm{EM}} + L_{\mathrm{e}}) = T_{\nu\mu} \tag{2.42}$$

$$T_{\mu\nu} = T_{\mathrm{EM}\mu\nu} + T_{\mathrm{Dirac}\mu\nu} \tag{2.43}$$

$$T_{EM\mu\nu} = -\frac{1}{4\pi} g^{\rho\sigma} F_{\mu\rho} F_{\nu\sigma} - g_{\mu\nu} L_{EM} = T_{EM\nu\mu} \tag{2.44}$$

$$T_{\mathrm{Dirac}\mu\nu} = -\varepsilon^{\Pi}{}_{\mu\nu} - \tau^{\Pi}{}_{\mu\nu}(g) - g_{\mu\nu} L_{\mathrm{Dirac}} = T_{\mathrm{Dirac}\nu\mu} \tag{2.45}$$

satisfies the conservation law

$$T^{\lambda}{}_{\mu;\lambda} = 0 \tag{2.46}$$

Also the antisymmetric angular momentum tensor

$$M^{\lambda\mu\nu} = x^{\mu} T^{\lambda\nu} - x^{\nu} T^{\lambda\mu} = -M^{\lambda\nu\mu} \tag{2.47}$$

satisfies the conservation law

$$\partial_{\lambda} M^{\lambda k\ell} = 0 \tag{2.48}$$

In Eq. (2.42), we have shown that the symmetric energy-momentum tensor $T_{\mu\nu}$ comprises not only the symmetric tensors but also polarized geometrical tensor $\varepsilon^{\Pi}{}_{\mu\nu}$ defined as

$$\begin{aligned}\varepsilon^{\Pi}{}_{\mu\nu} &= \frac{\hbar c}{4} e_{\lambda\nu} K^{*}{}_{\rho\sigma\mu} \varepsilon^{\lambda\rho\sigma\kappa} \overline{\Psi} \gamma_{\kappa} \gamma_{5} \Psi \\ &+ 2\left(\left(D^{*}{}_{\lambda} + T^{*}{}^{\kappa}_{\cdot}\kappa\lambda\right) F_{\mu\nu\cdot}{}^{\lambda} + T^{*}{}_{\rho\sigma\mu} F^{\rho}_{\cdot}{}^{\sigma}_{\cdot}\nu - \frac{1}{2} T^{*}{}_{\nu\rho\sigma} F^{\cdot\rho\sigma}_{\mu} \right)\end{aligned} \tag{2.49}$$

with

$$F^{abc} = \frac{\hbar c}{8} \varepsilon^{dabc} \overline{\Psi} \gamma_{d} \gamma_{5} \Psi \tag{2.50}$$

and polarized stress tensor $\tau^{\Pi}{}_{\mu\nu}(g)$ with the covariant derivative $D_{\mu}(g)$ given in Eq. (2.30):

$$\tau^{\Pi}{}_{\mu\nu}(g) = \frac{c}{2} \left(\overline{\psi} \gamma_{\nu} \left(-i\hbar D_{\mu}(g) \right) \psi + h.c. \right) \tag{2.51}$$

In this variation principle, *due to the presence of the spin connection* $\gamma_{ab\mu}$, *a new symmetry-polarized geometrical tensor* $\varepsilon^{\Pi}{}_{\mu\nu}$ *appears and whose antisymmetric component cancels with that of* $\tau^{\Pi}{}_{\mu\nu}(g)$ *as follows* (see Fig. 2.9):

$$\varepsilon^{A\mu\nu} + \tau^{A\mu\nu}(g) = 0 \tag{2.52}$$

where

$$\varepsilon^{\Pi\mu\nu} = \varepsilon^{S\mu\nu} + \varepsilon^{A\mu\nu} \tag{2.53}$$

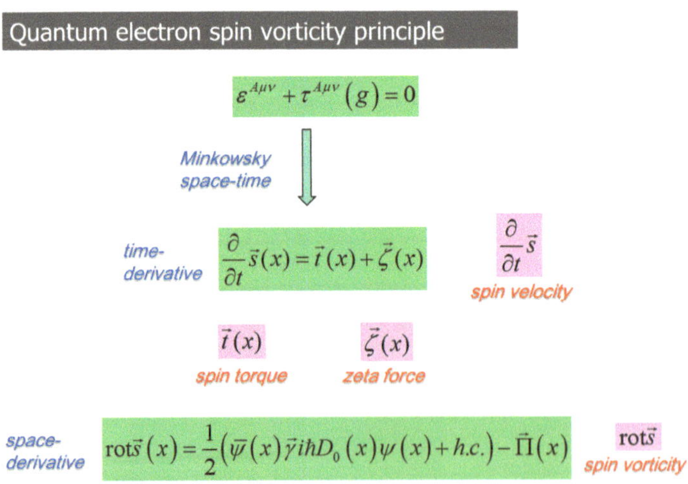

Fig. 2.9 Covariant form of the quantum electron spin vorticity principle

$$\varepsilon^{S\mu\nu} = \frac{1}{2}\left(\varepsilon^{\Pi\mu\nu} + \varepsilon^{\Pi\nu\mu}\right) \tag{2.54}$$

$$\varepsilon^{A\mu\nu} = \frac{1}{2}\left(\varepsilon^{\Pi\mu\nu} - \varepsilon^{\Pi\nu\mu}\right) \tag{2.55}$$

and

$$\tau^{\Pi\mu\nu}(g) = \tau^{S\mu\nu}(g) + \tau^{A\mu\nu}(g) \tag{2.56}$$

$$\tau^{S\mu\nu}(g) = \frac{1}{2}\left(\tau^{\Pi\mu\nu}(g) + \tau^{\Pi\nu\mu}(g)\right) \tag{2.57}$$

$$\tau^{A\mu\nu}(g) = \frac{1}{2}\left(\tau^{\Pi\mu\nu}(g) - \tau^{\Pi\nu\mu}(g)\right) \tag{2.58}$$

2.1.2 *The Minkowski Space-Time*

2.1.2.1 Spin Vorticity Principle

In the limit to the Minkowski space-time

$$e^{a}{}_{\mu} \longrightarrow \delta^{a}{}_{\mu}, \quad g_{\mu\nu} \longrightarrow \eta_{\mu\nu} \tag{2.59}$$

the equation of motion of the Dirac spinor field $\psi(x)$ is reduced from Eq. (2.37) to the Dirac Eq. (1.16) in due course. What is the physical meaning of Eq. (2.52)? The answer is twofold as is found if we take the limit to the Minkowski space-time.

First, for the time sector with $\mu = 0$, $\nu = 1, 2, 3$, we obtain

$$\mathrm{rot}\,\widehat{\vec{s}} + \widehat{\vec{\Pi}} - \frac{1}{2}\left(\widehat{\bar{\psi}}\vec{\gamma}\left(i\hbar\widehat{D}_0\right)\widehat{\psi} + h.c.\right) = 0 \tag{2.60}$$

Second, for the space sector with $\mu, \nu = 1, 2, 3$, we obtain

$$\frac{\partial}{\partial t}\widehat{\vec{s}} - \widehat{\vec{t}} - \widehat{\vec{\zeta}} = 0 \tag{2.61}$$

with torque $\widehat{\vec{t}}$ and zeta force $\widehat{\vec{\zeta}}$.

This Eq. (2.61) leads to the conclusion that the electron spin torque is counterbalanced by the chiral electron density, as found in Sect. 1.2.7, Chap. 1. The physical meaning of Eq. (2.52) is shown in Fig. 2.9.

2.1.2.2 Energy Density

In the limit to the Minkowski space-time, the energy-momentum tensor $T_{\text{Dirac}\mu\nu}$ is reduced to

$$T_{\text{Dirac}}{}^{\mu\nu} \to \begin{pmatrix} \frac{1}{2}(M+h.c.) & c\left(\vec{\Pi}+\frac{1}{2}\text{rot}\,\vec{s}\right)_x & c\left(\vec{\Pi}+\frac{1}{2}\text{rot}\,\vec{s}\right)_y & c\left(\vec{\Pi}+\frac{1}{2}\text{rot}\,\vec{s}\right)_z \\ c\left(\vec{\Pi}+\frac{1}{2}\text{rot}\,\vec{s}\right)_x & -\tau^S{}_{xx}+L_{\text{Dirac}} & -\tau^S{}_{xy} & -\tau^S{}_{xz} \\ c\left(\vec{\Pi}+\frac{1}{2}\text{rot}\,\vec{s}\right)_y & -\tau^S{}_{yx} & -\tau^S{}_{yy}+L_{\text{Dirac}} & -\tau^S{}_{yz} \\ c\left(\vec{\Pi}+\frac{1}{2}\text{rot}\,\vec{s}\right)_z & -\tau^S{}_{zx} & -\tau^S{}_{zy} & -\tau^S{}_{zz}+L_{\text{Dirac}} \end{pmatrix} \tag{2.62}$$

with the mass term M

$$M(x) = c\overline{\psi}(x)\left(-i\hbar\gamma^k D_k(x)+mc\right)\psi(x) \tag{2.63}$$

The electromagnetic component $T_{\text{EM}\mu\nu}$ of the energy-momentum tensor is also reduced to

$$T_{\text{EM}}{}^{\mu\nu} \to \begin{pmatrix} H_\gamma & cG_x & cG_y & cG_z \\ cG_x & \sigma_{xx} & \sigma_{xy} & \sigma_{xz} \\ cG_y & \sigma_{yx} & \sigma_{yy} & \sigma_{yz} \\ cG_z & \sigma_{zx} & \sigma_{zy} & \sigma_{zz} \end{pmatrix} \tag{2.64}$$

with the Poynting vector $\vec{G}$ and the Maxwell stress tensor $\overleftrightarrow{\sigma}$. The conservation law Eq. (2.46) of energy and momentum is then reduced to (see Figs. 2.10 and 2.11)

$$T^{\lambda 0}{}_{;\lambda} = 0 \to \frac{\partial}{\partial t}cP^0 + c^2\text{div}\vec{P} = 0 \tag{2.65}$$

$$T^{\lambda k}{}_{;\lambda} = 0 \to \frac{\partial}{\partial t}\vec{P} + \text{div}\left(\overleftrightarrow{\sigma} - \overleftrightarrow{\tau}^S\right) = 0 \tag{2.66}$$

$$P^\mu = \left(\frac{\frac{1}{2}(M+h.c.)+H_\gamma}{c}, \vec{\Pi}+\frac{1}{2}\text{rot}\,\vec{s}+\vec{G}\right) \tag{2.67}$$

It should be noted first that the vorticity plays an important role as momentum, and it is associated with antisymmetric electronic stress tensor $\overleftrightarrow{\tau}^A$. We may further prove that symmetric electronic stress tensor $\overleftrightarrow{\tau}^S$ plays an important role as tension $\vec{\tau}^S = \text{div}\overleftrightarrow{\tau}^S$ compensating the Lorentz force $\vec{L}$ as

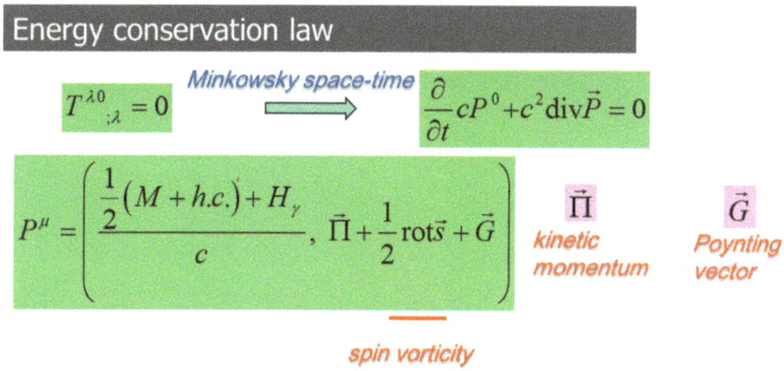

Fig. 2.10 Energy conservation law in the limit to the Minkowski space-time

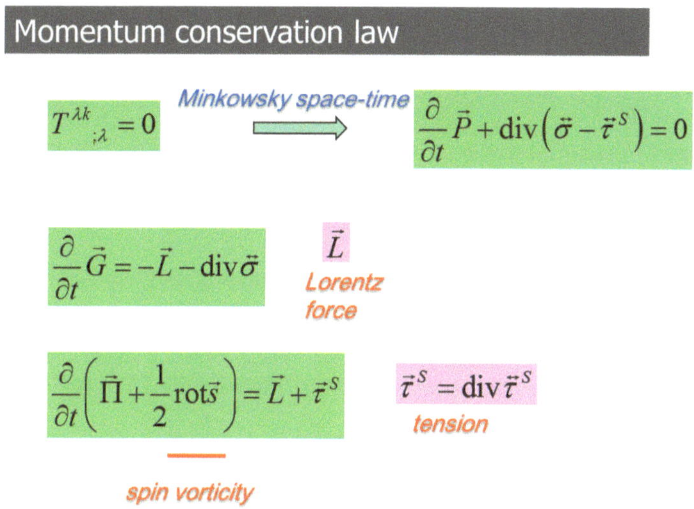

Fig. 2.11 Momentum conservation law in the limit to the Minkowski space-time

$$\frac{\partial}{\partial t} \vec{G} = -\vec{L} - \mathrm{div}\, \overleftrightarrow{\sigma} \tag{2.68}$$

$$P_{\mathrm{Dirac}} = \vec{\Pi} + \frac{1}{2}\mathrm{rot}\vec{s} \tag{2.69}$$

Consequently, after some manipulations, we finally arrive at

$$\frac{\partial}{\partial t} P_{\mathrm{Dirac}} = \frac{\partial}{\partial t}\left(\vec{\Pi} + \frac{1}{2}\mathrm{rot}\vec{s} \right) = \vec{L} + \vec{\tau}^S \tag{2.70}$$

$$\vec{\tau}^S = \mathrm{div}\overleftrightarrow{\tau}^S, \quad \vec{\tau}^{Sk} = \partial_\ell \tau^{Sk\ell} \tag{2.71}$$

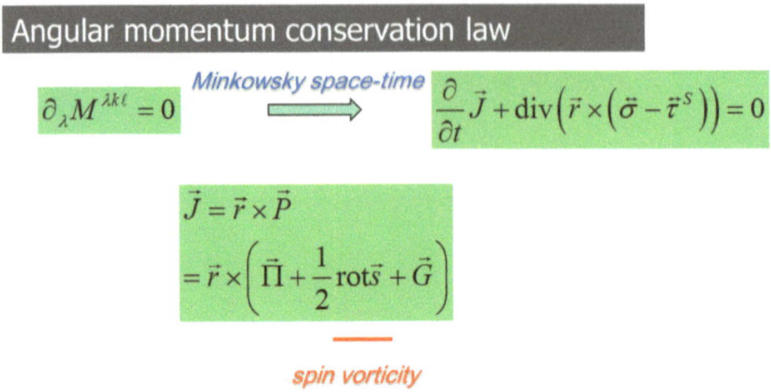

Fig. 2.12 Angular momentum conservation law in the limit to the Minkowski space-time

$$\tau^{S\mu\nu} = \frac{1}{2}\left(\tau^{\Pi\mu\nu} + \tau^{\Pi\nu\mu}\right) \tag{2.72}$$

This assures the equation of motion using solely the symmetric part of the tensor $\tau^S{}_{k\ell}$ *in the right-hand side.* This is the quantum electron spin vorticity principle in Sect. 1.2.8, Chap. 1 (see Fig. 1.22). The physical meaning of Eq. (2.52) is shown in Fig. 2.9.

Second, the conservation law Eq. (2.48) of angular momentum is then reduced to (see Fig. 2.12)

$$\partial_\lambda M^{\lambda k\ell} = 0 \quad \rightarrow \quad \frac{\partial}{\partial t}\vec{J} + \mathrm{div}\left(\vec{r}\times\left(\overleftrightarrow{\sigma} - \overleftrightarrow{\tau}^S\right)\right) = 0 \tag{2.73}$$

$$\frac{1}{c}M^{0k\ell} \quad \rightarrow \quad \vec{J} = \vec{r}\times\vec{\Pi} + \vec{r}\times\frac{1}{2}\mathrm{rot}\vec{s} + \vec{r}\times\vec{G} \tag{2.74}$$

Finally, for finite systems, the virial theorem is invoked (Landau and Lifshitz 1973) to arrive at

$$\int d^3\vec{r}\langle T^{00}\rangle \rightarrow E_{\mathrm{QED}} = mc^2\int d^3\vec{r}\langle\overline{\psi}\psi\rangle \tag{2.75}$$

We have shown the spin torque intrinsic to the spin-1/2 fermion is controlled by the chiral electron density, and the origin of the chiral nature is manifest in the principle of equivalence in general relativity. The time evolution of the electron spin is driven by the antisymmetric component of the electronic stress tensor through the vorticity. This is referred to as the quantum electron spin vorticity principle.

It is the relativity theory of spin, where inherent spin-orbit coupling is realized in between spin and orbital angular momentum. Imagine a bulk magnet. The magnet is phenomenologically composed of many magnetic domains. Each magnetic domain has its unique spin, which is the average of spin density within the domain. For the sake of simplicity, let first electron spin density respond to an applied magnetic field (or even an applied electric field in some cases in recent spintronics) and change through the domain wall, which is called "spin torque transfer" in the experiments of spintronics industry (note the simplification). Our prediction is that the spin torque does exist even in the stationary state when the spin torque is exactly canceled out with the zeta force. In the nonstationary state, however, the external magnetoelectric medium disturbs the intrinsic balance in between the spin torque and the zeta force established in the stationary state.

Of course, realistically, in addition to the spin of the electrons, the complexity origin of the magnetic spin can be either from the motion of electrons or nuclei, where the spin torque combinations totally can be treated by the equations of motion of angular momentum augmented by the ad hoc nuclear spin or more fundamentally the quark spin with the non-Abelian gauge.

For future technology of spintronics and photonics, the interaction of chirality of electron spin with another particle such as electron, nucleus, and photon (vector potential) should play an important role. Furthermore, the general relativity has recently been of vital importance with our daily life in particular for ultrahigh-precision communication with artificial satellite (e.g., GPS). The intrinsic formulations and the concrete analytical examples of the spin torque and zeta force presented in this book should help us understand the importance of chirality in modeling of materials of technological importance.

2.2 Rigged Field Theory

In application to chemical reaction dynamics, we have the Rigged QED theory where nuclear degrees of freedom are treated in a unified manner with QED. We shall examine here the nonrelativistic treatment of the Rigged QED theory and call it the primary Rigged QED theory. Note that in the nonrelativistic limit, we have a similar treatment by Lepage (Caswell and Lepage 1986).

2.2.1 Rigged QED Theory

Since we plug in the nuclear fields, we may first distinguish electron by the obvious suffix "e" in such a way as

$$\left(i\hbar\gamma^a e_a{}^\mu D_{e\mu}(g) - m_e c\right)\psi_e = 0 \tag{2.76}$$

We conventionally put the Schrödinger field equation of a'th nucleus onto the curved space-time. The procedures are (1) first, ignore the spin connection in $D_{e\mu}(g)$, (2) second, use the Dirac representation with ψ_e and approximate the small component as the multiplication of $-\frac{1}{2m_e c} i\hbar\sigma^k e_k{}^\mu D_{e\mu}$ to the large component, and (3) third, ignore again the spin-dependent terms in the resulting equation

$$\left(\sigma^k e_k{}^\mu D_{e\mu}\right)\left(\sigma^\ell e_\ell{}^\nu D_{e\nu}\right) \to \left(e_k{}^\mu D_{e\mu}\right)^2 \tag{2.77}$$

leading to

$$\left(i\hbar e_0{}^\mu D_{a\mu} - m_a c\right)\psi_a = \frac{(i\hbar)^2}{2m_a c}\left(e_k{}^\mu D_{a\mu}\right)^2 \psi_a \tag{2.78}$$

where the large component for electron is here used as ψ_a for the nuclear Schrödinger field. Note that the mass term is indispensable since we need it for the source of gravitation. We may identify this as the Schrödinger field equation without a priori spin and use this to plug in nuclear degrees of freedom into our formalism and call it as the Rigged QED theory in the curved space-time. In the course to the Minkowski space-time limit, this equation reduces to the usual Schrödinger field equation plus gravitational potential $m_a\Phi$ as shown in Sect. 2.4.1.

In the limit to the Minkowski space-time, we use the Dirac field Eq. (1.16) with the Schrödinger field Eq. (2.178) of a'th nucleus as follows:

$$\left(i\hbar\gamma^\mu D_{e\mu} - m_e c\right)\psi_e = 0 \tag{2.79}$$

$$i\hbar\frac{\partial}{\partial t}\chi_a = -\frac{\hbar^2}{2m_a}\vec{D}_a^{\,2}\chi_a + q_a A_0 \chi_a \tag{2.80}$$

where the gravitational potential $m_a\Phi$ in Eq. (2.178) is neglected (see Sect. 2.4.1). We have electron spin vorticity here but no spin vorticity for nuclei.

We get with obvious notation the momentum conservation law

$$\frac{\partial}{\partial t}\vec{P}_{\text{Rigged QED}} = -\text{div}\left(\overleftrightarrow{\sigma} - \overleftrightarrow{\tau}_{\text{Rigged QED}}{}^S\right) \tag{2.81}$$

and the angular momentum conservation law

$$\frac{\partial}{\partial t}\vec{J}_{\text{Rigged QED}} = -\text{div}\left(\vec{r} \times \left(\overleftrightarrow{\sigma} - \overleftrightarrow{\tau}_{\text{Rigged QED}}{}^S\right)\right) \tag{2.82}$$

The virial theorem for finite-system energetics is again used to obtain the energetics of the Rigged QED theory

$$E_{\text{Rigged QED}} = E_{\text{e;Rigged QED}} + \sum_a E_{a;\text{Rigged QED}} \tag{2.83}$$

$$E_{\text{e;Rigged QED}} = m_e c^2 \int d^3\vec{r} \langle \overline{\psi}_e \psi_e \rangle \tag{2.84}$$

$$E_{a;\text{Rigged QED}} = -\int d^3\vec{r} \langle T_a \rangle \tag{2.85}$$

$$T_a = -\frac{\hbar^2}{2m_a} \cdot \frac{1}{2}\left(\chi_a^\dagger \vec{D}_a^2 \chi_a + h.c.\right) \tag{2.86}$$

2.2.2 Primary Rigged QED Theory

We make approximation to electron as with Eq. (2.78) using symbols $\alpha = \mathrm{e}, a$ collectively as

$$\left(i\hbar e_0{}^\mu D_{\alpha\mu} - m_\alpha c\right)\psi_\alpha = \frac{(i\hbar)^2}{2m_\alpha c}\left(e_k{}^\mu D_{\alpha\mu}\right)^2 \psi_\alpha \tag{2.87}$$

In the limit to the Minkowski space-time, we further use approximation as of Eq. (2.80):

$$i\hbar \frac{\partial}{\partial t}\chi_\alpha = -\frac{\hbar^2}{2m_\alpha}\vec{D}_\alpha{}^2 \chi_\alpha + q_\alpha A_0 \chi_\alpha \tag{2.88}$$

We have lost the spin vorticity of electron and lost the antisymmetric component of the stress tensor of electron. We have the momentum conservation law

$$\frac{\partial}{\partial t}\vec{P}_{\text{Primary Rigged QED}} = -\mathrm{div}\left(\overleftrightarrow{\sigma} - \overleftrightarrow{\tau}_{\text{Primary Rigged QED}}{}^S\right) \tag{2.89}$$

and the angular momentum conservation law

$$\frac{\partial}{\partial t}\vec{J}_{\text{Primary Rigged QED}} = -\mathrm{div}\left(\vec{r} \times \left(\overleftrightarrow{\sigma} - \overleftrightarrow{\tau}_{\text{Primary Rigged QED}}{}^S\right)\right) \tag{2.90}$$

Using the virial theorem for finite-system energetics again, Eqs. (2.83), (2.84), (2.85), and (2.86) are reduced to the energetics of the primary Rigged QED theory

$$E_{\text{Primary Rigged QED}} = \sum_\alpha E_{\alpha;\text{ Primary Rigged QED}} \tag{2.91}$$

$$E_{\alpha;\ \text{Primary Rigged QED}} = \frac{1}{2}\int d^3\vec{r}\left\langle \tau_{\alpha;\ \text{Primary Rigged QED}}{}^{Skk}\right\rangle = -\int d^3\vec{r}\langle T_\alpha\rangle \quad (2.92)$$

$$\tau_{\alpha;\ \text{Primary Rigged QED}}{}^{Sk\ell} = \frac{\hbar^2}{4m_\alpha}\left(\chi_\alpha^\dagger D_{\alpha k}D_{\alpha l}\chi_\alpha - (D_{\alpha k}\chi_\alpha)^\dagger D_{\alpha l}\chi_\alpha + h.c.\right) \quad (2.93)$$

$$T_\alpha = -\frac{\hbar^2}{2m_\alpha}\cdot\frac{1}{2}\left(\chi_\alpha^\dagger \vec{D}_\alpha^2\chi_\alpha + h.c.\right) \quad (2.94)$$

2.3 SUGRA Energy-Momentum Tensor

2.3.1 *Stress Tensor*

In Sect. 2.1.1.2, the Dirac spinor field is a coordinate scalar and a Lorentz spinor, and the covariant derivative $D_\mu(g)$ is not only a coordinate scalar but also a Lorentz vector. It should be noted that the spin connection in the tetrad formalism is not unique. In SUGRA (see Sect. 2.3.3 for mathematical details), we have a new term $\gamma_{ab\mu}$(SUGRA) added to $\gamma_{ab\mu}$ as (Tachibana 2014)

$$\begin{aligned} D_\mu(\text{SUGRA}) &= \partial_\mu + i\frac{q}{\hbar c}A_\mu + i\frac{1}{2\hbar}\gamma_{ab\mu}J^{ab} + i\frac{1}{2\hbar}\gamma_{ab\mu}(\text{SUGRA})J^{ab} \\ &= D_\mu(g) + i\frac{1}{2\hbar}\gamma_{ab\mu}(\text{SUGRA})J^{ab} \end{aligned} \quad (2.95)$$

Then the symmetry-polarized stress tensor of electron $\tau^{\Pi}{}_{\mu\nu}(g)$ is changed to $\tau^{\Pi}{}_{\mu\nu}$(SUGRA) with the covariant derivative D_μ(SUGRA)

$$\tau^{\Pi}{}_{\mu\nu}(\text{SUGRA}) = \frac{c}{2}\left(\overline{\psi}\gamma_\nu(-i\hbar D_\mu(\text{SUGRA}))\psi + h.c.\right) \quad (2.96)$$

With the new spin connection term given, the new symmetry-polarized geometrical tensor $\varepsilon^{\Pi\mu\nu}$(SUGRA) *appears, and again now that the energy-momentum tensor* $T_{\mu\nu}$(SUGRA) *is symmetric and hence the electronic part* $T_{e\mu\nu}$(SUGRA) *is symmetric, the resultant antisymmetric component of the* $\varepsilon^{A\mu\nu}$(SUGRA) *cancels with* $\tau^{A\mu\nu}$ (SUGRA):

$$\varepsilon^{A\mu\nu}(\text{SUGRA}) + \tau^{A\mu\nu}(\text{SUGRA}) = 0 \quad (2.97)$$

where

$$\varepsilon^{A\mu\nu}(\text{SUGRA}) = \frac{1}{2}\left(\varepsilon^{\Pi\mu\nu}(\text{SUGRA}) - \varepsilon^{\Pi\nu\mu}(\text{SUGRA})\right) \quad (2.98)$$

$$\tau^{A\mu\nu}(\text{SUGRA}) = \frac{1}{2}\left(\tau^{\Pi\mu\nu}(\text{SUGRA}) - \tau^{\Pi\nu\mu}(\text{SUGRA})\right) \tag{2.99}$$

2.3.2 Energy-Momentum Tensor

We shall examine an example of the symmetric energy-momentum tensor of a simple SUGRA in the case of a simple SUSY with linearized gravity. See Sect. 2.3.3 as mathematical Appendix.

A weak classical gravity is represented by the infinitesimal transformation (Weinberg 1995)

$$x^\mu(x) \to x'^\mu(x) = x^\mu(x) + \xi^\mu(x) \tag{2.100}$$

$$\Lambda^a{}_b(x) \to \Lambda'^a{}_b(x) = \delta^a{}_b + \omega^a{}_b(x) \tag{2.101}$$

$$e^a{}_\mu(x) = \delta^a{}_\mu + 2k\phi^a{}_\mu(x) \to e'^a{}_\mu(x') = \delta^a{}_\mu + 2k\phi'^a{}_\mu(x') \tag{2.102}$$

$$\phi_{\mu\nu}(x) \to \phi'_{\mu\nu}(x') = \phi_{\mu\nu}(x) + \frac{1}{2k}\left(-\frac{\partial\xi_\mu(x)}{\partial x^\nu} + \omega_{\mu\nu}(x)\right) \tag{2.103}$$

where

$$k = \sqrt{8\pi G}\frac{\hbar}{c^2} \tag{2.104}$$

This leads to a weak gravitational field $h_{\mu\nu}(x)$ as

$$g_{\mu\nu}(x) = \eta_{\mu\nu} + 2kh_{\mu\nu}(x) \tag{2.105}$$

$$h_{\mu\nu}(x) = \phi_{\mu\nu}(x) + \phi_{\nu\mu}(x) \tag{2.106}$$

The action integral given in Eq. (2.1) is cast into the linearized form as

$$\begin{aligned} I &= \frac{c}{2\kappa}\int R\sqrt{-g}d^4x + \frac{1}{c}\int L\sqrt{-g}d^4x \\ &\xrightarrow{\text{linearized}} I_{\text{linearized}} = \frac{1}{c}\int d^4x\left(-\hbar^2 E^{\mu\nu}h_{\mu\nu} - kT^{\mu\nu}h_{\mu\nu} + L^{(0)}{}_{\text{linearized}}\right) \end{aligned} \tag{2.107}$$

$$\begin{aligned} G_{\text{linearized}}{}^{\mu\nu} &= 2kE^{\mu\nu}, \quad E^{\mu\nu} \\ &= \frac{1}{2}\begin{pmatrix} \Box h^{\mu\nu} - \partial^\alpha\partial^\mu h_\alpha{}^\nu - \partial^\alpha\partial^\nu h_\alpha{}^\mu \\ +\partial^\nu\partial^\mu h^\alpha{}_\alpha - \eta^{\mu\nu}\Box h^\alpha{}_\alpha + \eta^{\mu\nu}\partial^\alpha\partial^\beta h_{\alpha\beta} \end{pmatrix} = E^{\nu\mu} \end{aligned} \tag{2.108}$$

where $E^{\mu\nu}$ is the linearized Einstein tensor and $L^{(0)}{}_{\text{linearized}}$ is the linearized Lagrangian density of QED excluding the interaction with gravity. *In the right-hand side of*

Eq. (2.107), we have the symmetric energy-momentum tensor $T_{\mu\nu} = T_{\nu\mu}$ *and hence the symmetric stress tensor* $T_{e\mu\nu} = T_{e\nu\mu}$ *as the electronic part.*

In SUGRA, we have the gauge transformation of the spin-2 field of graviton $h_{\mu\nu}(x)$ as

$$h_{\mu\nu}(x) \to h'_{\mu\nu}(x) = h_{\mu\nu}(x) - \frac{1}{2k}\left(\frac{\partial \xi_\mu(x)}{\partial x^\nu} + \frac{\partial \xi_\nu(x)}{\partial x^\mu}\right) \tag{2.109}$$

The graviton is associated with the superpartner, called the gravitino $\psi_\mu(x)$, represented by the spin-3/2 Rarita–Schwinger field, whose gauge transformation is

$$\psi_\mu(x) \to \psi'_\mu(x) = \psi_\mu(x) - \partial_\mu \psi(x) \tag{2.110}$$

where $\psi(x)$ is a spin-1/2 Majorana field. These are the components of the metric superfield $H_\mu(x)$, whose gauge transformation is

$$H_\mu(x) \to H'_\mu(x) = H_\mu(x) - \Delta_\mu(x) \tag{2.111}$$

where $\Delta_\mu(x)$ is given by the linear superfield $\bar{\mathcal{D}}\Xi(x)$ as

$$\Delta_\mu(x) = \bar{\mathcal{D}}\Xi(x)\gamma_\mu \tag{2.112}$$

The gauge fields are then calculated to be

$$\phi_{\mu\nu}(x) = V^H{}_{\mu\nu}(x) - \frac{1}{3}\eta_{\mu\nu} V^{H\lambda}{}_\lambda(x) \tag{2.113}$$

$$\psi_\mu(x) = 2\lambda^H{}_\mu(x) - \frac{2}{3}\gamma_\mu \not{\lambda}^H(x) + \frac{2}{3} i\hbar\gamma_\mu \partial^\rho \omega^H{}_\rho(x) \tag{2.114}$$

with

$$\xi_\mu(x) = 2kv_\mu(x) \tag{2.115}$$

$$v_\mu(x) = -\hbar\overline{\omega}^\Xi(x)\gamma^\mu + \text{const} \tag{2.116}$$

$$\omega_{\mu\nu}(x) = k\big(\partial_\nu v_\mu(x) - \partial_\mu v_\nu(x) - V^\Delta{}_{\mu\nu}(x) + V^\Delta{}_{\nu\mu}(x)\big) \tag{2.117}$$

$$\psi(x) = 4i\hbar M^\Xi(x) - 4\hbar\gamma_5 N^\Xi(x) + \text{const} \tag{2.118}$$

Consequently, the gauge-invariant linearized SUGRA action integral is found to be

$$I_{\text{linearized}}(\text{SUGRA}) = I_{\text{linearized}} - \frac{1}{c}\int d^4x \begin{pmatrix} \frac{1}{2}\overline{\Psi}_\mu L^\mu \times c - \frac{1}{2}k\hbar^{-1}\bar{S}_{new}{}^\mu \psi_\mu - \frac{4}{3}(b^\mu b_\mu + 2p^2 + 2s^2) \\ +k\hbar^{-1}(-R^\mu b_\mu + 2pA^X - 2sB^X) \end{pmatrix} \tag{2.119}$$

where $S_{\text{new}}{}^\mu$ is the supersymmetry current,

$$R^\mu = 2C^{\Theta\mu} \tag{2.120}$$

is the$R$-current, and the others are

$$L^\sigma = -\frac{\hbar}{c}\varepsilon^{\nu\mu\kappa\sigma}\partial_\kappa \gamma_5 \gamma_\nu \psi_\mu \tag{2.121}$$

$$b^\sigma = D^{H\sigma} - \hbar^2 \partial^\sigma \partial_\mu C^{H\mu} + \frac{1}{2}\hbar\varepsilon^{\nu\mu\kappa\sigma}\partial_\kappa V^H{}_{\mu\nu} \tag{2.122}$$

$$p = i\hbar\partial^\mu N^H{}_\mu \tag{2.123}$$

$$s = i\hbar\partial^\mu M^H{}_\mu \tag{2.124}$$

Further optimization of the auxiliary fields b^μ ,p, and s leads to

$$I_{\text{linearized}_{opt}}(\text{SUGRA}) = I_{\text{linearized}} - \frac{1}{c}\int d^4x \begin{pmatrix} \frac{1}{2}\overline{\Psi}_\mu L^\mu \times c - \frac{1}{2}k\hbar^{-1}\bar{S}_{new}{}^\mu \psi_\mu \\ +\frac{3}{8}k^2\hbar^{-2}\left(\frac{1}{2}R^\mu R_\mu + (A^X)^2 + (B^X)^2\right) \end{pmatrix} \tag{2.125}$$

We may identify the negative energy density $-\frac{3}{8}k^2\hbar^{-2}\left((A^X)^2 + (B^X)^2\right)$ for the anti-de Sitter space-time.

$$\rho_{VAC}(x) = \rho_{SVAC}(x) - \frac{3}{8}k^2\hbar^{-2}\left((A^X(x))^2 + (B^X(x))^2\right) \tag{2.126}$$

We have the SUGRA action added to$I_{\text{linearized}}$, *as shown in Eqs. (2.119) and (2.125), so that we have again the symmetric energy-momentum tensor* $T_{\mu\nu} = T_{\nu\mu}$ *and hence the symmetric stress tensor* $T_{e\mu\nu} = T_{e\nu\mu}$ *as the electronic part.*

2.3.3 SUGRA Formalism

2.3.3.1 The Majorana Spinor

First, the Majorana spinor satisfies

$$\mathrm{C}\theta = -\theta \tag{2.127}$$

$$\overline{\theta} = \theta^\dagger \gamma^0 = -{}^t\theta C = {}^t\theta' {}^tC = {}^t(C\theta) \tag{2.128}$$

$${}^t\left(\frac{\partial}{\partial \overline{\theta}}\right)(\overline{\theta} M \theta) = 2M\theta \tag{2.129}$$

$$\overline{\theta}_1 M \theta_2 = \overline{\theta}_2 C^{-1t} {}^tM C \theta_1 \tag{2.130}$$

$$\theta\overline{\theta} = -\frac{1}{4}(\overline{\theta}\theta) + \frac{1}{4}\gamma_\mu \gamma^5 (\overline{\theta}\gamma_5 \gamma^\mu \theta) + \frac{1}{4}\gamma^5 (\overline{\theta}\gamma_5 \theta) \tag{2.131}$$

A spinor is decomposed into a pair of the Majorana spinors as

$$s = \theta_+ + i\theta_- \tag{2.132}$$

$$\theta_+ = \frac{1}{2}(1 - \mathbf{C})\mathrm{s}, \quad \theta_- = \frac{1}{2i}(1 + \mathbf{C})\mathrm{s} \tag{2.133}$$

$$\mathbf{C}\theta_\pm = -\theta_\pm \tag{2.134}$$

2.3.3.2 The Haag–Lopuszanski–Sohnius Theorem

The $\left(0, \frac{1}{2}\right)$-fermionic generator $Q_{\dot{U}\,r}$ is transformed under the Lorentz transformation as

$$U(\Lambda^{-1}) Q_{\dot{U}\,r} U(\Lambda) = \lambda_{\eta \dot{U}}{}^{\dot{V}}(\Lambda) Q_{\dot{V}\,r} \tag{2.135}$$

$$\left[\vec{J}, Q_{\dot{U}r}\right] = -\frac{\hbar}{2}\vec{\sigma}_{\dot{U}}{}^{\dot{V}} Q_{\dot{V}\,r}, \quad \left[\vec{K}, Q_{\dot{U}r}\right] = -\frac{\hbar}{2} i\vec{\sigma}_{\dot{U}}{}^{\dot{V}} Q_{\dot{V}\,r} \tag{2.136}$$

$$\left[\vec{A}, Q_{\dot{U}r}\right] = 0, \quad \left[\vec{B}, Q_{\dot{U}r}\right] = -\frac{\hbar}{2}\vec{\sigma}_{\dot{U}}{}^{\dot{V}} Q_{\dot{V}\,r} \tag{2.137}$$

with the charge conjugation operator $\mathbf{C}$ and the complex conjugate operator K, the Dirac spinor representation is

$$\mathbf{C}\begin{pmatrix} e^{AW} K Q_{\dot{W}\,r} \\ Q_{\dot{U}\,r} \end{pmatrix}\mathbf{C}^{-1} = -\begin{pmatrix} e^{AW} K Q_{\dot{W}\,r} \\ Q_{\dot{U}\,r} \end{pmatrix} \tag{2.138}$$

Likewise, the $\left(\frac{1}{2}, 0\right)$-fermionic generator P^{As} is transformed under the Lorentz transformation as

$$U(\Lambda^{-1}) P^{As} U(\Lambda) = \lambda_\xi{}^A{}_B(\Lambda) P^{Bs} \tag{2.139}$$

$$\left[\vec{J}, P^{As}\right] = -\frac{\hbar}{2}\vec{\sigma}_B^A P^{Bs}, \quad \left[\vec{K}, P^{As}\right] = +\frac{\hbar}{2} i\vec{\sigma}_B^A P^{Bs} \tag{2.140}$$

$$\left[\vec{A}, P^{As}\right] = -\frac{\hbar}{2}\vec{\sigma}_B^A P^{Bs}, \quad \left[\vec{B}, P^{As}\right] = 0 \tag{2.141}$$

with Dirac spinor representation

$$\mathbf{C}\begin{pmatrix} P^{As} \\ -e_{\dot{U}\dot{B}} K P^{Bs} \end{pmatrix}\mathbf{C}^{-1} = -\begin{pmatrix} P^{As} \\ -e_{\dot{U}\dot{B}} K P^{Bs} \end{pmatrix} \tag{2.142}$$

The Haag–Lopuszanski–Sohnius theorem states that

$$\left\{Q_{\dot{U}r}, K Q_{\dot{V}s}\right\} = 2\delta_{rs}(\sigma^\mu)_{\dot{U}V} P_\mu \tag{2.143}$$

$$\left\{Q_{\dot{U}r}, Q_{\dot{V}s}\right\} = e_{\dot{U}\dot{V}} Z_{rs}, \quad Z_{rs} = -Z_{sr} \tag{2.144}$$

$$\left\{\begin{pmatrix} e^{AW} K Q_{\dot{W}r} \\ Q_{\dot{U}r} \end{pmatrix}, \begin{pmatrix} e^{BX} K Q_{\dot{X}s} \\ Q_{\dot{U}s} \end{pmatrix}^\dagger \gamma_0\right\} = 2\gamma^\mu P_\mu \delta_{rs} - \frac{1+\gamma_5}{2} K Z_{rs} + \frac{1-\gamma_5}{2} Z_{rs} \tag{2.145}$$

where P_μ is the 4-momentum operator and Z_{rs} are the central charges. For simple supersymmetry, we have null Z_{rs}.

2.3.3.3 The Salam–Strathdee Superfield with Simple SUSY

The Salam–Strathdee superfield S with simple SUSY is constructed by using the Majorana spinors θ, ω, and λ as

$$\begin{aligned} S = {} & C - i(\overline{\theta}\gamma_5\omega) \\ & -\frac{1}{2} i(\overline{\theta}\gamma_5\theta)M - \frac{1}{2}(\overline{\theta}\theta)N - \frac{1}{2}(\overline{\theta}\gamma_5\gamma_\mu\theta)V^\mu \\ & -i(\overline{\theta}\gamma_5\theta)\left(\overline{\theta}\left(\lambda - \frac{1}{2} i\hbar\partial\!\!\!/\omega\right)\right) - \frac{1}{4}(\overline{\theta}\gamma_5\theta)^2\left(D - \frac{1}{2}\hbar^2\Box C\right) \end{aligned} \tag{2.146}$$

where the component C of S may be emphasized with superscript C^S, etc. Taking the *h.c.*, we have

$$\begin{aligned} S^\dagger = {} & C^\dagger - i(\overline{\theta}\gamma_5\omega) \\ & -\frac{1}{2} i(\overline{\theta}\gamma_5\theta)M^\dagger - \frac{1}{2}(\overline{\theta}\theta)N^\dagger - \frac{1}{2}(\overline{\theta}\gamma_5\gamma_\mu\theta)(V^\mu)^\dagger \\ & -i(\overline{\theta}\gamma_5\theta)\left(\overline{\theta}\left(\lambda - \frac{1}{2} i\hbar\partial\!\!\!/\omega\right)\right) - \frac{1}{4}(\overline{\theta}\gamma_5\theta)^2\left(D^\dagger - \frac{1}{2}\hbar^2\Box C^\dagger\right) \end{aligned} \tag{2.147}$$

If with the Hermitian superfield $S^\dagger = S$, we have

$$C^\dagger, M^\dagger, N^\dagger, V^{\mu\dagger}, D^\dagger = C, M, N, V^\mu, D \tag{2.148}$$

The infinitesimal translation δS is defined as

$$\begin{aligned}\delta S &= \frac{1}{i}\left(\overline{\alpha}\begin{pmatrix} e^{AW} K\mathcal{Q}_{\dot{W}\,r} \\ \mathcal{Q}_{\dot{U}\,r}\end{pmatrix}\right)S \\ &= (\overline{\alpha}\mathcal{Q})S\end{aligned} \tag{2.149}$$

$$\left[\begin{pmatrix} e^{AW} K\mathcal{Q}_{\dot{W}\,r} \\ \mathcal{Q}_{\dot{U}\,r}\end{pmatrix}, S\right\} = i\mathcal{Q}S \tag{2.150}$$

The generator Q should then satisfy

$$\{\mathcal{Q}_{\dot{U}}, {}^t(C\mathcal{Q})V\} = -2i\hbar(\sigma^\mu)_{\dot{U}\,V}\partial_\mu \tag{2.151}$$

$$\{\mathcal{Q}_{\dot{U}}, \mathcal{Q}_{\dot{V}}\} = 0 \tag{2.152}$$

with the Dirac spinor representation

$$\{\mathcal{Q}_\ell, \bar{\mathcal{Q}}_{\bar{\ell}}\} = -2i\hbar(\gamma^\mu)_{\ell\bar{\ell}}\partial_\mu \tag{2.153}$$

Also, the generator in the superfield coordinate representation is given as

$$\mathcal{Q} = -{}^t\left(\frac{\partial}{\partial\bar{\theta}}\right) - i\hbar\partial\!\!\!/\theta = {}^tC\frac{\partial}{\partial\theta} - i\hbar\partial\!\!\!/\theta \tag{2.154}$$

$$\bar{\mathcal{Q}} = {}^t(Cq) = {}^t\left(\frac{\partial}{\partial\theta}\right) + i\hbar\bar{\theta}\partial\!\!\!/ \tag{2.155}$$

with the Dirac spinor representation

$$\mathcal{Q}_\ell = C_{\bar{\ell}\ell}\frac{\partial}{\partial\theta_{\bar{\ell}}} - i\hbar(\gamma^\mu)_{\ell\bar{\ell}}\theta_{\bar{\ell}}\partial_\mu \tag{2.156}$$

$$\bar{\mathcal{Q}}_\ell = \mathcal{Q}_{\bar{\ell}}C_{\bar{\ell}\ell} = {}^t\left(\frac{\partial}{\partial\theta_\ell}\right) - i\hbar{}^t(C\gamma^\mu\theta)\ell\partial_\mu \tag{2.157}$$

The derivative in the superfield coordinate representation is defined as

$$\mathcal{D} = -{}^t\left(\frac{\partial}{\partial\bar{\theta}}\right) + i\hbar\partial\!\!\!/\theta = {}^tC\frac{\partial}{\partial\theta} + i\hbar\partial\!\!\!/\theta \tag{2.158}$$

$$\bar{\mathcal{D}} = {}^t(C\mathcal{D}) = {}^t\left(\frac{\partial}{\partial\theta}\right) - i\hbar\bar{\theta}\partial\!\!\!/ \tag{2.159}$$

with the Dirac spinor representation

$$\{\mathcal{D}_\ell, \bar{\mathcal{D}}_{\bar{\ell}}\} = 2i\hbar(\gamma^\mu)_{\ell\bar{\ell}}\partial_\mu \tag{2.160}$$

We have the commutation relationships

$$\{\mathcal{D}, \mathcal{Q}\} = 0 \tag{2.161}$$

and

$$[\mathcal{D}, \delta] = 0 \tag{2.162}$$

The infinitesimal translation of the components of S should then be obtained as

$$\delta C = i(\overline{\alpha}\gamma_5\omega) \tag{2.163}$$

$$\delta\omega = \left(-\hbar\gamma_5\partial\!\!\!/ C - M + i\gamma_5 N + iV\!\!\!/\right)\alpha \tag{2.164}$$

$$\delta M = -\left(\overline{\alpha}\left(\lambda - i\hbar\partial\!\!\!/\omega\right)\right) \tag{2.165}$$

$$\delta N = i\left(\overline{\alpha}\gamma_5\left(\lambda - i\hbar\partial\!\!\!/\omega\right)\right) \tag{2.166}$$

$$\delta V^\mu = (\overline{\alpha}\gamma^\mu\lambda) - i\hbar(\overline{\alpha}\partial^\mu\omega) \tag{2.167}$$

$$\delta\lambda = \left(\frac{1}{2}\hbar\left[\partial_\mu V\!\!\!/, \gamma^\mu\right] + i\gamma_5 D\right)\alpha \tag{2.168}$$

$$\delta D = i\left(\overline{\alpha}\gamma_5\left(-i\hbar\partial\!\!\!/\lambda\right)\right) \tag{2.169}$$

Action integral for interaction of supercurrent and metric superfield is introduced as

$$I_{\text{int}} = 2k\int d^4x\left[H_\mu(x)\Theta^\mu(x)\right]_D/\hbar c \tag{2.170}$$

The supercurrent is defined as

$$\Theta_\mu = \frac{i}{12}\sum_n\left(-4\hbar^2\left({\Phi_n}^\dagger\partial_\mu\Phi_n - \Phi_n\partial_\mu{\Phi_n}^\dagger\right) - i\hbar\left(\left(\bar{\mathcal{D}}{\Phi_n}^\dagger\right)\gamma_\mu\left(\mathcal{D}\Phi_n\right)\right)\right) \times c \tag{2.171}$$

where Φ_n is the chiral superfield

$$\begin{aligned}\Phi_n = \phi_n - \sqrt{2}\left(\overline{\theta}\frac{1+\gamma_5}{2}\psi_n\right) + \left(\overline{\theta}\frac{1+\gamma_5}{2}\theta\right)F_n - \frac{1}{2}i\hbar\left(\overline{\theta}\gamma_5\partial\!\!\!/\phi_n\theta\right)\\ + \frac{1}{\sqrt{2}}i\hbar\left(\overline{\theta}\gamma_5\theta\right)\left(\overline{\theta}\frac{1-\gamma_5}{2}\partial\!\!\!/\psi_n\right) + \frac{1}{8}\hbar^2\left(\overline{\theta}\gamma_5\theta\right)^2\Box\phi_n\end{aligned} \tag{2.172}$$

The supercurrent conservation laws are found to be

$$\gamma^{\mu}\mathcal{D}\Theta_{\mu} = \mathcal{D}X \tag{2.173}$$

$$\bar{\mathcal{D}}\Theta_{\mu}\gamma^{\mu} = -\bar{\mathcal{D}}X \tag{2.174}$$

where X denotes the real chiral superfield

$$\begin{aligned} X = &A - (\overline{\theta}\psi) \\ &-\frac{1}{2}i(\overline{\theta}\gamma_5\theta)G + \frac{1}{2}(\overline{\theta}\theta)F + \frac{1}{2}\hbar(\overline{\theta}\gamma_5\gamma_{\mu}\theta)\partial^{\mu}B \\ &+(\overline{\theta}\gamma_5\theta)\left(\overline{\theta}\left(-\frac{1}{2}i\hbar\gamma_5\partial\!\!\!/\psi\right)\right) + \frac{1}{8}\hbar^2(\overline{\theta}\gamma_5\theta)^2\Box A \end{aligned} \tag{2.175}$$

2.4 Examples

2.4.1 Rigged QED Theory in the Curved Space-Time

Put

$$\psi_a = \chi_a e^{-im_a c^2 t/\hbar} \tag{2.176}$$

in Eq. (2.78) under the weak gravitation condition with only non-Kronecker delta insertion

$$e_0{}^0 \sim \frac{1}{\sqrt{1 + 2\frac{\Phi}{c^2}}} \sim 1 - \frac{\Phi}{c^2} \tag{2.177}$$

Then we obtain

$$i\hbar\frac{\partial}{\partial t}\chi_a = \left(-\frac{\hbar^2}{2m_a}\vec{D}_a{}^2 + q_a A_0 + m_a\Phi\right)\chi_a \tag{2.178}$$

This is the correct equation of motion for neutron, if we identify Φ as the gravitation potential

$$i\hbar\frac{\partial}{\partial t}\chi_{\text{neutron}} = \left(-\frac{\hbar^2}{2m_{\text{neutron}}}\Delta + m_{\text{neutron}}\Phi\right)\chi_{\text{neutron}} \tag{2.179}$$

That Eq. (2.179) is correct has been demonstrated experimentally using a neutron interferometer (Collela et al. 1975).

2.4.2 The Majorana Particle

The fundamental equations of motion of the Majorana particle are summarized.

The Majorana equations are

$$\left(i\hbar\partial^{\nu}(\sigma_{\nu})^{A\dot{U}} \pm m_{L}e^{i\delta_{L}}ce^{AU}K\right)\eta_{\dot{U}} = 0 \tag{2.180}$$

$$(i\hbar\partial_{\nu}(\sigma^{\nu})_{\dot{U}A} \mp m_{R}e^{i\delta_{R}}ce_{\dot{U}\dot{A}}K)\xi^{A} = 0 \tag{2.181}$$

with the Klein–Gordon equations

$$\left((i\hbar\partial)^{2} - (m_{L}c)^{2}\right)\eta_{\dot{U}} = 0 \tag{2.182}$$

$$\left((i\hbar\partial)^{2} - (m_{R}c)^{2}\right)\xi^{A} = 0 \tag{2.183}$$

where $m_{R,L}$ are the real masses and $\delta_{R,L}$ are the real phases.

The charge conjugation properties are

$$\mathbf{C}|\psi_{\mathrm{M}_1}\rangle = \xi_{\mathrm{M}_1}|\psi_{\mathrm{M}_1{}^c}\rangle = |\psi_{\mathrm{M}_1}\rangle \tag{2.184}$$

$$\psi_{\mathrm{M}_1{}^c} = C\gamma^{0}K\psi_{\mathrm{M}_1} = -\psi_{\mathrm{M}_1} \tag{2.185}$$

$$\mathbf{C}|\psi_{\mathrm{M}_2}\rangle = \xi_{\mathrm{M}_2}|\psi_{\mathrm{M}_2{}^c}\rangle = |\psi_{\mathrm{M}_2}\rangle \tag{2.186}$$

$$\psi_{\mathrm{M}_2{}^c} = C\gamma^{0}K\psi_{\mathrm{M}_2} = -\psi_{\mathrm{M}_2} \tag{2.187}$$

$$\xi_{\mathrm{M}_1} = \xi_{\mathrm{M}_2} = -1 \tag{2.188}$$

The Dirac spinor representations are

$$(i\hbar\partial\!\!\!/ \pm m_{L}e^{i\delta_{L}}c)\psi_{\mathrm{M}_1} = 0, \quad \psi_{\mathrm{M}_1} = \begin{pmatrix} e^{AW}K\eta_{\dot{W}} \\ \eta_{\dot{U}} \end{pmatrix} \tag{2.189}$$

$$(i\hbar\partial\!\!\!/ \pm m_{R}e^{i\delta_{R}}c)\psi_{\mathrm{M}_2} = 0, \quad \psi_{\mathrm{M}_2} = \begin{pmatrix} \xi^{A} \\ -e_{\dot{U}\dot{B}}K\xi^{B} \end{pmatrix} \tag{2.190}$$

and

$$i\hbar\partial\!\!\!/\Psi_{\mathrm{M}} \pm m_{\mathrm{M}}c(-)\Psi_{\mathrm{M}^c} = 0 \tag{2.191}$$

$$\Psi_{\mathrm{M}} = \begin{pmatrix} \xi^{A} \\ \eta_{\dot{U}} \end{pmatrix} \tag{2.192}$$

$$\Psi_{\mathrm{M}^c} = C\gamma^{0}K\Psi_{\mathrm{M}} = (-)\begin{pmatrix} e^{AW}K\eta_{\dot{W}} \\ -e_{\dot{U}\dot{B}}K\xi^{B} \end{pmatrix} \tag{2.193}$$

$$m_{\mathrm{M}}c = \begin{pmatrix} m_L e^{i\delta_L} c & 0 \\ 0 & m_R e^{i\delta_R} c \end{pmatrix} \tag{2.194}$$

2.4.3 The Atiyah–Singer Index Theorem

We quickly review the radiation corrections to the currents of electron in QED.

The charge is conserved but not the chiral charge since we have no continuity equation for the latter because of the nonzero mass of electron. Actually, we have residual pseudoscalar as the fourth-rank antisymmetric tensor

$$\frac{1}{cq}\partial_\mu j_5{}^\mu(x) = i\frac{2m_\mathrm{e}c}{\hbar}\overline{\psi}(x)\gamma_5\psi(x) \tag{2.195}$$

which is not zero unless m_e is zero.

The Euclidean path integrals have been widely used to treat the radiative corrections, where we realize the corrected charge current $J^\mu(x)$ satisfies the conservation law

$$\partial_\mu J^\mu(x) = 0 \tag{2.196}$$

by the Ward–Takahashi identities, but Eq. (2.195) is modified for the corrected chiral current $J_5{}^\mu(x)$

$$\begin{aligned} &\left\langle \tfrac{1}{cq}\partial_\mu J_5{}^\mu(x)\right\rangle_H\Big|_{A \text{ fixed}} - i\frac{2m_\mathrm{e}c}{\hbar}\left\langle\overline{\Psi}(x)\gamma_5\Psi(x)\right\rangle_H\Big|_{A \text{ fixed}} \\ &= 2\times\frac{1}{32\pi^2}\left(\frac{q}{\hbar c}\right)^2\varepsilon^{\mu\nu\rho\sigma}F_{\mu\nu}(x)F_{\rho\sigma}(x) \end{aligned} \tag{2.197}$$

under the fixed background field $A^\mu(x)$ as

$$\langle O_H(y)\rangle_H\big|_{A \text{ fixed}} = \left.\frac{\displaystyle\int[d\psi(x)][d\overline{\psi}(x)]\exp\left[\frac{i}{\hbar c}\int d^4xL(x)\right]o(y)}{\displaystyle\int[d\psi(x)][d\overline{\psi}(x)]\exp\left[\frac{i}{\hbar c}\int d^4xL(x)\right]}\right|_{A \text{ fixed}} \tag{2.198}$$

The right-hand side of Eq. (2.197) is the Chern–Pontryagin density

$$A(x) = \frac{1}{32\pi^2}\left(\frac{q}{\hbar c}\right)^2\varepsilon^{\mu\nu\rho\sigma}F_{\mu\nu}(x)F_{\rho\sigma}(x) \tag{2.199}$$

This is known as the axial anomaly and is proportional to the divergence of the Chern–Simons class $G_\mu(x)$:

$$G_\mu(x) = \varepsilon_{\mu\nu\rho\sigma} A^\nu(x) \partial^\rho A^\sigma(x) \tag{2.200}$$

$$\partial_\mu G^\mu(x) = \frac{1}{2} F_{\mu\nu}(x) \widetilde{F}^{\mu\nu}(x) \tag{2.201}$$

with

$$\widetilde{F}^{\mu\nu}(x) = \frac{1}{2} \varepsilon^{\mu\nu\rho\sigma} F_{\rho\sigma}(x) \tag{2.202}$$

The topology of the gauge field obeys the Atiyah–Singer index theorem:

$$\begin{aligned} \nu &= \int d^4x A(x) \\ &= \int d^4x \frac{1}{32\pi^2} \left(\frac{q}{\hbar c}\right)^2 \varepsilon^{\mu\nu\rho\sigma} F_{\mu\nu}(x) F_{\rho\sigma}(x) \end{aligned} \tag{2.203}$$

Here ν is the index of $i\hbar \not{D}(x)$:

$$\nu = \mathrm{ind}(i\hbar \not{D}(x)) = n_+ - n_- = \int d^4x \sum_{n_0} \overline{\phi}^{(\pm)}{}_{n_0}(x) \gamma_5 \phi^{(\pm)}{}_{n_0}(x) \tag{2.204}$$

$$i\hbar \not{D}(x) \phi^{(\pm)}{}_{n_0}(x) = 0 \phi^{(\pm)}{}_{n_0}(x) \tag{2.205}$$

$$\gamma_5 \phi^{(\pm)}{}_{n_0}(x) = \pm \phi^{(\pm)}{}_{n_0}(x) \tag{2.206}$$

where $n_\pm$ are the number of zero modes of $i\hbar \not{D}(x)$ that have eigenvalues ± 1 for γ_5.

2.5 Summary

QED is reformulated in a way that is covariant under general coordinate transformation. The consequence gives the right answer to the odd question "what is momentum of electron spin?" raised in Chap. 1. The whole picture of stress tensor in Chap. 1 has thus been unified.

In application to chemical reaction dynamics in finite systems, we have devised the Rigged QED theory where nuclear degrees of freedom are treated in a unified manner with QED. The nonrelativistic treatment of the Rigged QED theory is also examined as the primary Rigged QED theory. We have used the virial theorem for the finite-system energetics of the Rigged QED theory and the primary Rigged QED theory.

The theory has also been extended in this chapter using a simple SUGRA, which is a simple SUSY model of gravity.

References

W.E. Caswell, G.P. Lepage, Phys. Lett. B **167**, 437 (1986)
R. Collela, A.W. Overhauser, S.A. Werner, Phys. Rev. Lett. **34**, 1472 (1975)
R. Haag, M. Sohnius, J.T. Łopuszański, Nucl. Phys. B **88**, 257 (1975)
K. Hayashi, T. Shirafuji, Phys. Rev. D **19**, 3524 (1979)
F.W. Hehl, P. von der Heyde, G.D. Kerlick, J.M. Nester, Rev. Mod. Phys. **48**, 393 (1976)
L.D. Landau, E.M. Lifshitz, *The Classical Theory of Fields*, 6th ed. (Nauka, Moscow, 1973), translated into English in *Course of Theoretical Physics Series, Volume* **2**, 4th edn. (Pergamon, Oxford, 1975)
A. Lopes, Nature **471**, 459 (2011)
N. Nakanishi, Prog. Theor. Phys. **111**, 301 (2004)
A. Tachibana, J. Math. Chem. **50**, 669 (2012)
A. Tachibana, Indian J. Chem. Sect. A **53**, 1031 (2014)
S. Weinberg, *Gravitation and Cosmology: Principles and Applications of the General Theory of Relativity* (Wiley, New York, 1972)
S. Weinberg, *The Quantum Theory of Fields I-III* (Cambridge University, Cambridge, 1995)

Chapter 3
Chemical Ideas of QED

Abstract In Chap. 1, we have studied the symmetry-polarized internal self-stress tensor of electron. In Chap. 2, we have studied the antisymmetric component of the stress tensor of electron in detail. It remains a simple question: for what does the symmetric component work? The answer is first chemical idea of QED is the spindle structure of covalency predicted by the symmetric component of the stress tensor of electron. We have so long considered relativistic theory as merely a slight correction for the interpretation in chemical phenomena. However, we shall clarify that the Hamiltonian of QED, derived from the picture of "action through medium" based on the relativistic theory, gives a novel image of the chemical interaction even in the nonrelativistic limit. Actually, we shall use the primary Rigged QED theory for our purpose. As a result, conventional images of the chemical interaction based on "action at a distance" are replaced with the new images of them given by the picture of "action through medium" without exception. We shall visualize not only the spindle structure but also various basic chemical concepts in chemical reaction systems in real three-dimensional space. Shape volume discriminates the region of classical atoms and molecules. The regional chemical potential inequality principle gives the electron transferability from one region to another.

Keywords Lagrange point • Lagrange surface • Nonclassical bond order • Primary Rigged QED theory • Regional chemical potential • Shape volume • Spindle structure • Stress tensor • Tension

3.1 Overview

Atomic theory of the universe coined by an ancient Greek pre-Socratic philosopher Democritus was reformulated by Einstein through his theory of the Brownian motion of minute particles suspended in a liquid and has been proved by Perrin experimentally (see Fig. 3.1). Pauling and Wilson have elaborated chemical bond theory of atoms using quantum chemistry (Pauling and Wilson 1935).

Atom as minimum unit of human body is a cell. As an analogy to the spindle structure observed in cell division (see Fig. 3.2), it is predicted by QED in covalent bond division (see Fig. 3.3).

A. Tachibana, *New Aspects of Quantum Electrodynamics*,
DOI 10.1007/978-981-10-3132-8_3

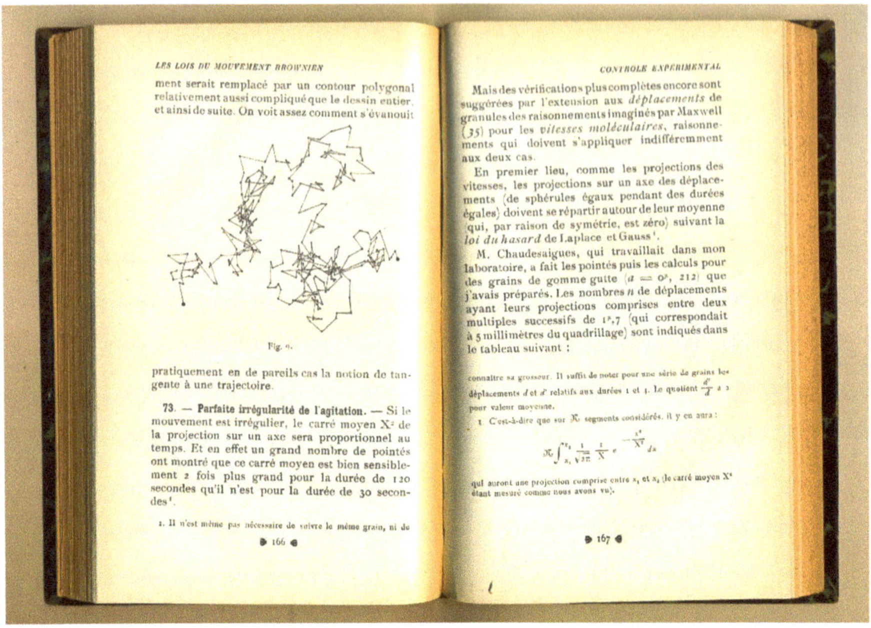

LES LOIS DU MOUVEMENT BROWNIEN

ment serait remplacé par un contour polygonal relativement aussi compliqué que le dessin entier, et ainsi de suite. On voit assez comment s'évanouit

Fig. 9.

pratiquement en de pareils cas la notion de tangente à une trajectoire.

73. — Parfaite irrégularité de l'agitation. — Si le mouvement est irrégulier, le carré moyen X^2 de la projection sur un axe sera proportionnel au temps. Et en effet un grand nombre de pointés ont montré que ce carré moyen est bien sensiblement 2 fois plus grand pour la durée de 120 secondes qu'il n'est pour la durée de 30 secondes[1].

1. Il n'est même pas nécessaire de suivre le même grain, ni de

166

CONTROLE EXPERIMENTAL

Mais des vérifications plus complètes encore sont suggérées par l'extension aux *déplacements* de granules des raisonnements imaginés par Maxwell (35) pour les *vitesses moléculaires*, raisonnements qui doivent s'appliquer indifféremment aux deux cas.

En premier lieu, comme les projections des vitesses, les projections sur un axe des déplacements (de sphérules égaux pendant des durées égales) doivent se répartir autour de leur moyenne (qui, par raison de symétrie, est zéro) suivant la *loi du hasard* de Laplace et Gauss[1].

M. Chaudesaigues, qui travaillait dans mon laboratoire, a fait les pointés puis les calculs pour des grains de gomme gutte ($a = 0^{\mu},212$) que j'avais préparés. Les nombres n de déplacements ayant leurs projections comprises entre deux multiples successifs de $1^{\mu},7$ (qui correspondait à 5 millimètres du quadrillage) sont indiqués dans le tableau suivant :

connaître sa grosseur. Il suffit de noter pour une série de grains les déplacements d et d' relatifs aux durées 1 et 4. Le quotient $\frac{d'}{d}$ a 2 pour valeur moyenne.

1. C'est-à-dire que sur $\mathfrak{N}$ segments considérés, il y en aura :

$$\mathfrak{N}\int_{x_1}^{x_2} \frac{1}{\sqrt{2\pi}} \frac{1}{X} e^{-\frac{x^2}{X^2}} dx$$

qui auront une projection comprise entre x_1 et x_2 (le carré moyen X^2 étant mesuré comme nous avons vu).

167

Fig. 3.1 Experimental proof of atom (Reproduced from Perrin 1914)

The key idea underlying the spindle structure is the *internal self-stress* of QED (see Chap. 1) as applied to study a unified scheme for generalized chemical reactivity. The chemical reactivity in this scheme is the force acting on a pair of electronic drop regions (Tachibana 2001, 2002). *This is a new kind of chemical force acting in between electrons not in between nuclei.* A new look at the chemical bond is thereby elucidated.

The spindle structure here in QED is a geometrical object of a region where principal electronic stress is positive along a line of principal axis of the electronic stress that connects a pair of the electronic drop region R_Ds of atoms and molecules. The spindle structure of covalency is the first chemical idea of QED that is alternative to the occupancy of bonding molecular orbital (see Fig. 3.4).

The anti-spindle structure of no covalency should also be characteristic of QED that is alternative to the overwhelming occupancy of antibonding molecular orbital over and above the bonding molecular orbital. The bonding and antibonding energy densities are visualized locally.

The concept of energy density using the stress tensor of QED is found in Sect. 3.1.2 and onward in this chapter for more details (Tachibana 2003, 2004). The symmetrical component $\overleftrightarrow{\tau}^{S}$ of the electronic stress tensor has been proved to predict the emergence of the covalent bond in terms of the spindle structure; see Sect. 3.2 and onward in this chapter for more details. The theory of the spindle structure has also been developed to visualize the nonclassical bond order concept of chemical bond and the regional chemical potential. For physicochemical properties of

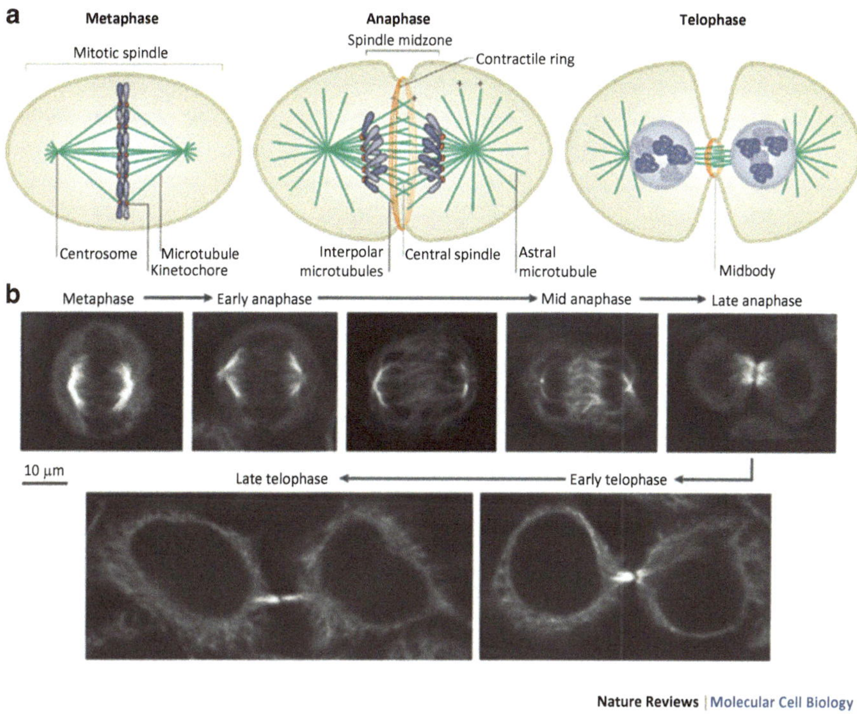

Fig. 3.2 Spindle structure of mitosis (Reproduced from Glotzer 2009)

materials, the Coulson conjecture of energy density (Coulson 1961) has been organized in terms of thermodynamics by using the regional chemical potential; see Sect. 3.5 and onward in this chapter for more details (Tachibana 2001).

In this chapter, all numerical calculations are shown in atomic unit using the nonrelativistic limit of the primary Rigged QED theory (see Sect. 2.2.2, Chap. 2) unless otherwise stated explicitly.

3.1.1 *Primary Rigged QED Theory*

We use the primary Rigged QED theory presented in Sect. 2.2.2, Chap. 2 as

$$\hat{H}_{\text{Primary Rigged QED}}(x) = \hat{H}_\gamma(x) + \sum_\alpha \hat{T}_\alpha(x) \tag{3.1}$$

$$\hat{T}_\alpha(x) = -\frac{\hbar^2}{2m_\alpha} \cdot \frac{1}{2}\left(\hat{\chi}^\dagger_\alpha(x)\hat{\vec{D}}^2_\alpha(x)\hat{\chi}_\alpha(x) + h.c.\right) \tag{3.2}$$

The energy flow is found to be

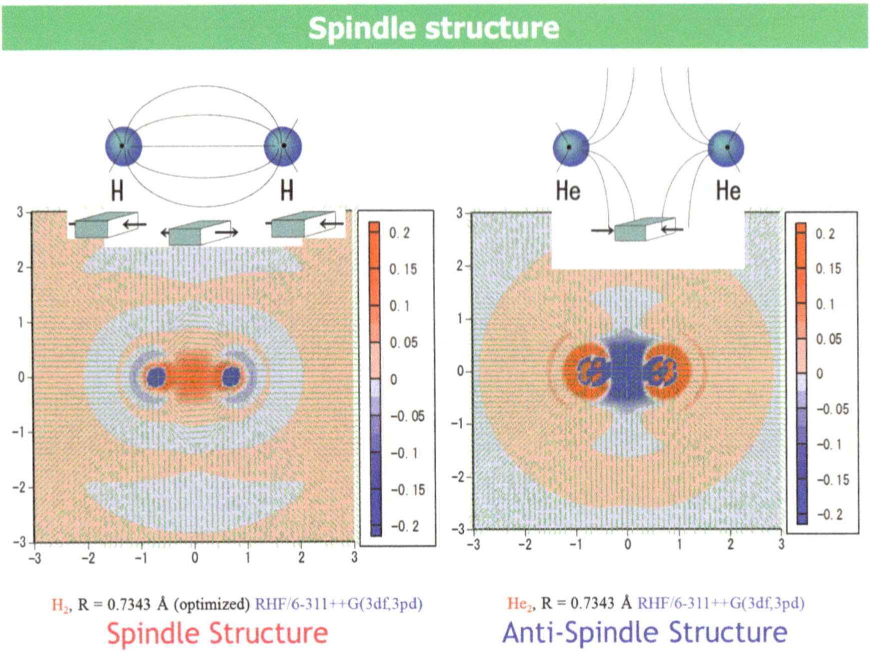

Fig. 3.3 Spindle structure for H_2 (*left*) and anti-spindle structure for He_2 (*right*)

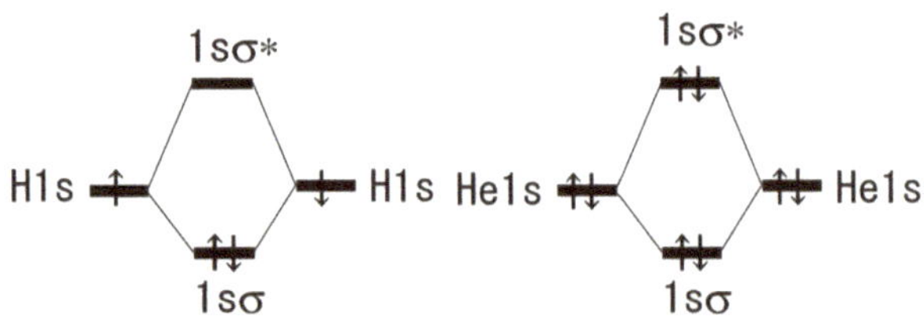

Fig. 3.4 Orbital interaction scheme of covalent bond (*left*) and no covalent bond (*right*)

$$\frac{\partial}{\partial t}\widehat{H}_{\text{Primary Rigged QED}}(x) = -\mathrm{div}\left(c^2\frac{1}{2}\left(\widehat{\vec{G}}(x)+\widehat{\vec{G}}^\dagger(x)\right)+\sum_\alpha \widehat{\vec{s}}_\alpha(x)\right) \tag{3.3}$$

with

$$\begin{aligned}\widehat{s}^k_\alpha(x) = \frac{1}{2i\hbar}\left(\frac{\hbar^2}{2m_\alpha}\right)\Big(&-\widehat{\chi}^\dagger_\alpha(x)\widehat{D}_{\alpha k}(x)\widehat{\vec{D}}^2_\alpha(x)\widehat{\chi}_\alpha(x)\\ &+\left(\widehat{D}_{\alpha k}(x)\widehat{\chi}_\alpha(x)\right)^\dagger\widehat{\vec{D}}^2_\alpha(x)\widehat{\chi}_\alpha(x) - h.c.\Big)\end{aligned} \tag{3.4}$$

The virial theorem of the primary Rigged QED theory leads to

$$E_{\text{Primary Rigged QED}} = \sum_{\alpha} E_{\alpha;\text{Primary Rigged QED}} \tag{3.5}$$

$$E_{\alpha;\text{Primary Rigged QED}} = \frac{1}{2}\int d^3\vec{r}\left\langle \hat{\tau}_{\alpha;\text{Primary Rigged QED}}{}^{Skk}(x)\right\rangle = -\int d^3\vec{r}\left\langle \hat{T}_{\alpha}(x)\right\rangle \tag{3.6}$$

where

$$\hat{\tau}_{\alpha;\text{Primary Rigged QED}}{}^{Sk\ell}(x) = \frac{\hbar^2}{4m_\alpha}\left(\hat{\chi}^{\dagger}_{\alpha}(x)\hat{D}_{\alpha k}(x)\hat{D}_{\alpha l}(x)\hat{\chi}_{\alpha}(x) - \left(\hat{D}_{\alpha k}(x)\hat{\chi}_{\alpha}(x)\right)^{\dagger}\hat{D}_{\alpha l}(x)\hat{\chi}_{\alpha}(x) + h.c.\right) \tag{3.7}$$

$$\left\langle \hat{\overleftrightarrow{\tau}}_{\alpha;\text{Primary Rigged QED}}{}^{S}(x)\right\rangle = \overleftrightarrow{\tau}^{S}_{\alpha}(x) = \begin{pmatrix} \tau^{S}_{\alpha xx}(x) & \tau^{S}_{\alpha xy}(x) & \tau^{S}_{\alpha xz}(x) \\ \tau^{S}_{\alpha yx}(x) & \tau^{S}_{\alpha yy}(x) & \tau^{S}_{\alpha yz}(x) \\ \tau^{S}_{\alpha zx}(x) & \tau^{S}_{\alpha zy}(x) & \tau^{S}_{\alpha zz}(x) \end{pmatrix}$$
$$\xrightarrow{\text{diag}} \begin{pmatrix} \tau^{S11}_{\alpha}(x) & 0 & 0 \\ 0 & \tau^{S22}_{\alpha}(x) & 0 \\ 0 & 0 & \tau^{S33}_{\alpha}(x) \end{pmatrix}, \quad \tau^{S11}_{\alpha}(x) \le \tau^{S22}_{\alpha}(x) \le \tau^{S33}_{\alpha}(x) \tag{3.8}$$

$$\hat{\vec{\tau}}^{S}{}_{\alpha}(x) = \text{div}\hat{\overleftrightarrow{\tau}}_{\alpha}{}^{S}(x) \tag{3.9}$$

$$\hat{T}_{\alpha}(x) = -\frac{\hbar^2}{2m_\alpha}\cdot\frac{1}{2}\left(\hat{\chi}^{\dagger}_{\alpha}(x)\hat{\vec{D}}^{2}_{\alpha}(x)\hat{\chi}_{\alpha}(x) + h.c.\right) \tag{3.10}$$

It should be noted that the nuclear motion is cast in the field theory. So the wave-packet prescription of the nuclear motion is indispensable for the simple picture of the regional energy partitioning Eq. (3.5). The adiabatic approximation gives rise to another virial for nuclei, which vanishes for equilibrium nuclear configurations but remains finite for nonequilibrium nuclear configurations.

3.1.2 *Shape Volume of Shell Structure and the Intrinsic Electronic Transition State*

The redistribution of electron is essential in the course of chemical reaction coordinate (Tachibana 1987b, 1996, 1999a). This is because the associated lowering in the electronic energy is the driving force of chemical reaction (Tachibana and Parr 1992). The decomposition of the electronic energy in the abstract functional space of orbital has played a significant role in the study of chemical reactivity indices such as the Coulson valence bond theory (Coulson 1961), the Woodward–Hoffmann law (Fleming 1976), and the Fukui frontier orbital theory (Fukui 1981). The pathway of the nuclear configuration change is given by using the theory of the intrinsic reaction coordinate (IRC) or meta-IRC

(Tachibana 1991, 1994) and the nuclear dynamics in terms of differential geometry (Tachibana and Iwai 1986; Tachibana 1999b) with the vibronic application to superconductivity (Tachibana 1987a).

We have recently developed a novel theory of energy decomposition in the real space (Tachibana 2002). The new energy decomposition scheme is exact and complementary to the conventional orbital space energy decomposition scheme. Namely, which region of space has significant contribution to chemical reaction coordinate is easily recognized. This is advantageous in visualization of the chemical interaction in real space. This new regional energy decomposition scheme has been extended to infinitely small regional energy decomposition scheme, namely, the electronic energy density decomposition scheme. Using the electronic energy density, we can pick up any point in a chemical reaction system and find how the electronic energy is assigned to the point. We can then integrate the electronic energy density in a small region and find out the regional electronic energy contribution to the global electronic energy. If the integration spans the whole space, then the integral gives the total.

Another look at the density of electron is the kinetic energy density. This is based on the observation of the Einstein equation

$$\left(\frac{E}{c}-\frac{q_\alpha}{c}\phi\right)^2-\left(\vec{p}-\frac{q_\alpha}{c}\vec{A}\right)^2=(m_\alpha c)^2 \tag{3.11}$$

which states that the square of the gauge-invariant 4-momentum $p^\mu - \frac{q_\alpha}{c}A^\mu$ should be invariant under the Lorentz transformation. The field theoretical version for electron reads

$$\begin{aligned}&\left\langle\hat{\psi}^\dagger(x)\left(\frac{1}{c}i\hbar\frac{\partial}{\partial t}-\frac{q_e}{c}\hat{\phi}(x)\right)^2\hat{\psi}(x)\right\rangle-\left\langle\hat{\psi}^\dagger(x)\left(-i\hbar\vec{\nabla}-\frac{q_e}{c}\hat{\vec{A}}(x)\right)^2\hat{\psi}(x)\right\rangle\\&=(m_e c)^2\langle\hat{\psi}^\dagger(x)\hat{\psi}(x)\rangle\end{aligned} \tag{3.12}$$

The positivity of Eq. (3.12) is the measure of classical reality since it is proportional to the kinetic energy density in the primary Rigged QED theory

$$n_{T_e}(x)=\left\langle\hat{T}_e(x)\right\rangle \tag{3.13}$$

The $n_{T_e}(\vec{r})$ is a measure of the shell structure. In the very vicinity of atomic nucleus, the electron feels infinitely large positive electric potential of the bare nucleus. Then, in terms of classical mechanics, the electron that has constant energy can acquire infinitely large positive kinetic energy at the position of the nucleus. In terms of quantum mechanics as well, the $n_{T_e}(\vec{r})$ should then become infinitely positive at the position of nucleus, provided that, which is the very case of normal chemistry, the intramolecular electric field $\vec{E}_{\mathrm{intra}}(\vec{r})$ produced by the other electrons does not exceed that of the bare nucleus (Kato 1957; Bingel 1963, 1967;

Fig. 3.5 Electronic drop and atmosphere regions

Electronic drop region

$R_D = \{\vec{r} \mid n_{T_e}(\vec{r}) > 0\}$ ⇒ Neoclassical reality

Electronic atmosphere region

$R_A = \{\vec{r} \mid n_{T_e}(\vec{r}) < 0\}$ ⇒ Quantum tunneling

Electronic interface

$S = \{\vec{r} \mid n_{T_e}(\vec{r}) = 0\}$

Pack and Brown 1966). The nucleus is therefore normally surrounded by the surface of zero kinetic energy density, $n_{T_e}(\vec{r}) = 0$, within which the kinetic energy density $n_{T_e}(\vec{r}) > 0$ where the electron density is amply accumulated and classically allowed motion of electron is guaranteed. Then, we may call this the region of the electronic drop denoted by R_D and the complementary region of the electronic atmosphere denoted by R_A, being separated by the electronic interface S (see Fig. 3.5)

$$R_D = \{\vec{r} | n_{T_e}(\vec{r}) > 0\}, \quad R_A = \{\vec{r} | n_{T_e}(\vec{r}) < 0\}, \quad S = \{\vec{r} | n_{T_e}(\vec{r}) = 0\} \tag{3.14}$$

Within R_D, electrons can move freely as in classical mechanics, whereas toward R_A, they can tunnel through S. Then S describes the union of turning points for electrons in R_D in the generic sense.

Here we first pick up hydrogen-like atom in the ground state for which we have nonrelativistic limit of the kinetic energy density $n_{T_e}(\vec{r})$ (Tachibana 2013)

$$n_{T_e}(\vec{r}) \xrightarrow{\text{nonrelativistic limit}} \frac{{Z_{\text{eff}}}^5 e^2}{\pi a^4}\left(\frac{2}{x} - \frac{1}{2}\right)e^{-x}, \quad x = \frac{2Z_{\text{eff}}}{a} r, \quad a = \frac{\hbar^2}{m_e e^2} \tag{3.15}$$

$$r_{S_{\text{Dirac}}} \text{ in the ground state} = \begin{cases} \dfrac{\hbar}{m_e c Z_{\text{eff}} \alpha}\left(\sqrt{1 - (Z_{\text{eff}}\alpha)^2} + \sqrt[4]{1 - (Z_{\text{eff}}\alpha)^2}\right) \text{(primary Rigged QED)} \\ \dfrac{\hbar}{m_e c Z_{\text{eff}} \alpha}\left(\sqrt{1 - (Z_{\text{eff}}\alpha)^2} + 1\right) \text{(4-component Dirac wave function)} \end{cases} \tag{3.16}$$

In the ground state, the nucleus is surrounded by the electronic interface of radius $r_{S_{\text{Dirac}}}$ for the relativistic case and r_S for the nonrelativistic case within which the kinetic energy density is positive leaving negative outside (see Fig. 3.6).

So the intrinsic shape is this electronic drop region just surrounded by this electronic interface. The radius for the relativistic case is smaller than the nonrelativistic value r_S, which ratio approaches 0 as Z approaches to 137. The $r_{S_{\text{Dirac}}}$

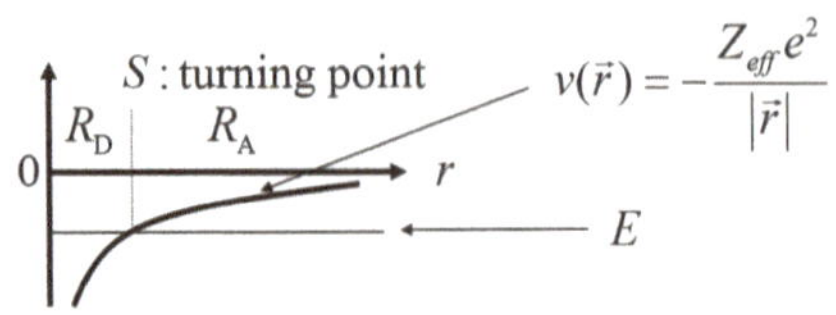

$r_{S_{Dirac}}$ in the ground state =

$$\begin{cases} \frac{\hbar}{m_e c Z_{eff}\alpha}\left(\sqrt{1-(Z_{eff}\alpha)^2} + \sqrt[4]{1-(Z_{eff}\alpha)^2}\right) \text{(Primary Rigged QED)} \\ \frac{\hbar}{m_e c Z_{eff}\alpha}\left(\sqrt{1-(Z_{eff}\alpha)^2} + 1\right) \text{(4-component Dirac wave function)} \end{cases}$$

$$\xrightarrow{\text{nonrelativistic limit}} r_S = 2\frac{\hbar}{m_e c Z_{eff}\alpha}$$

Fig. 3.6 Turning point for the shape volume of hydrogen atom in the ground state

intrinsic shape of H atom with $Z = 1$ has the radius $r_{S_{Dirac}} \approx r_S = 2$ [a.u.] ≈ 1.058Å, which is comparable to the standard atomic radius of H atom, 1.5 Bohr ≈ 0.794Å, reduced by a factor of ca. 75%.

In R_A the electron density is dried up and the motion of electron is classically forbidden. The boundary S in between R_D and R_A gives a clear image of the *intrinsic shape volume* of the *shell structure* in the reactant atoms and molecules, the reaction intermediates, and the reaction products along the course of the chemical reaction coordinate. In Fig. 3.7 is shown the R_D of $H + H \rightarrow H_2$ chemical reaction system from top with internuclear distance $R = 6.0$ Å to bottom 0.8 Å. Two initially disjoint R_Ds merge in between. The *intrinsic electronic transition state* is defined for $R = R^{\dagger}$ when two disjoint shape volumes of H atoms merge (Tachibana 2001): $R^{\dagger}$ is 4 Bohr $\approx$ 2.117 Å according to Eq. (3.76) in an analytical model.

3.2 Stress Tensor and the Spindle Structure

The symmetric stress tensor $\overleftrightarrow{\tau}_{\alpha}^{S}(\vec{r})$ in Eq. (3.8) gives the tensorial energy density. The eigenvalue of the symmetric stress tensor is the principal stress, and the eigenvector is the principal axis (see Fig. 3.8).

Stress tensors of one-dimensional stationary states with rectangular potentials and harmonic potentials have negative compressive stresses. In analogy with the classical sense, the particle has one-dimensional liquid character (see Fig. 3.9).

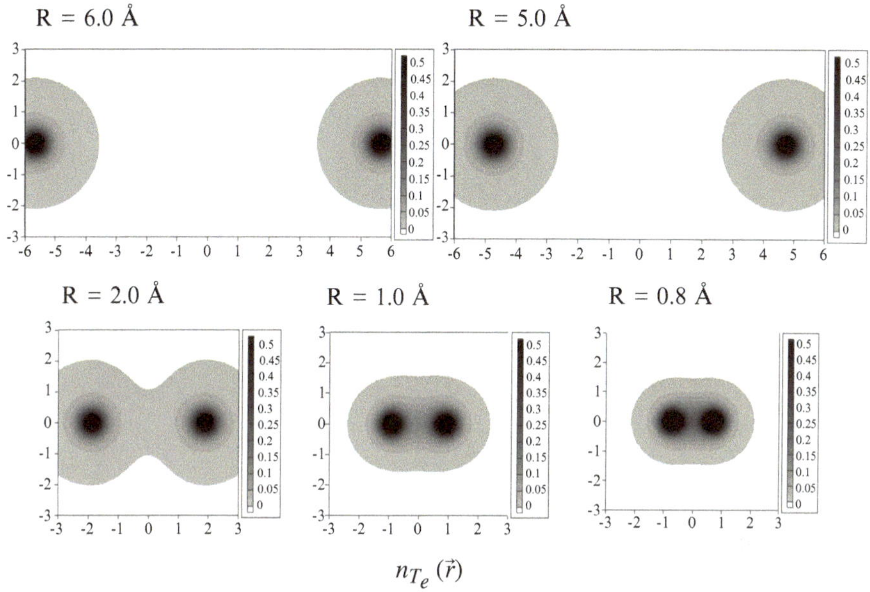

Fig. 3.7 Shape volume R_D along the reaction coordinate H+H → H_2 from *top left* with internuclear distance $R = 6.0$ Å to *bottom right* 0.8 Å using MRCI/6–311++G(3df,3pd)

Fig. 3.8 Principal stress and principal axis of the tensorial energy density

$$\vec{\tau}^{S}_{\alpha}(\vec{r}) = \begin{pmatrix} \tau^{S}_{\alpha xx}(\vec{r}) & \tau^{S}_{\alpha xy}(\vec{r}) & \tau^{S}_{\alpha xz}(\vec{r}) \\ \tau^{S}_{\alpha yx}(\vec{r}) & \tau^{S}_{\alpha yy}(\vec{r}) & \tau^{S}_{\alpha yz}(\vec{r}) \\ \tau^{S}_{\alpha zx}(\vec{r}) & \tau^{S}_{\alpha zy}(\vec{r}) & \tau^{S}_{\alpha zz}(\vec{r}) \end{pmatrix}$$

$$\xrightarrow{\text{diag}} \begin{pmatrix} \tau^{S11}_{\alpha}(\vec{r}) & 0 & 0 \\ 0 & \tau^{S22}_{\alpha}(\vec{r}) & 0 \\ 0 & 0 & \tau^{S33}_{\alpha}(\vec{r}) \end{pmatrix}$$

$$\tau^{S11}_{\alpha}(\vec{r}) \le \tau^{S22}_{\alpha}(\vec{r}) \le \tau^{S33}_{\alpha}(\vec{r})$$

The particle may be bound in a one-dimensional potential energy box or quantum harmonic oscillator or scattering as a superposition of incident, reflected, or transmitted component of the stationary wave.

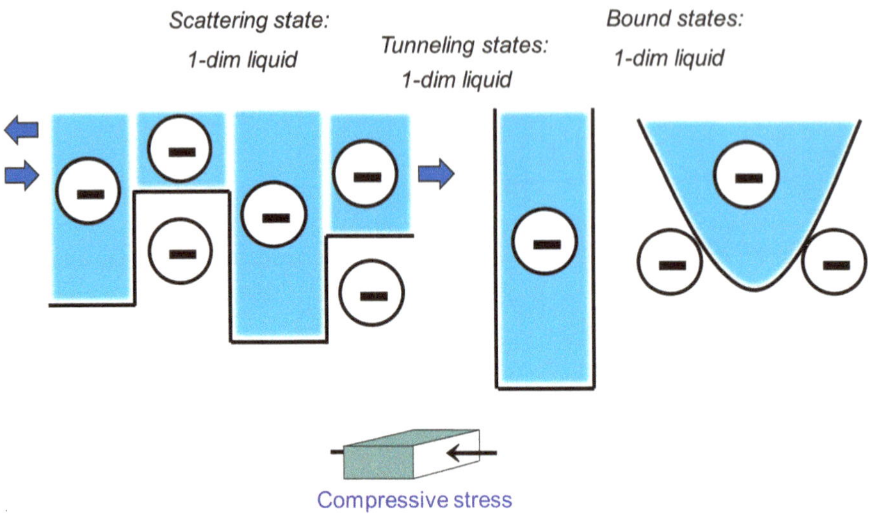

Fig. 3.9 Stationary one-dimensional liquids

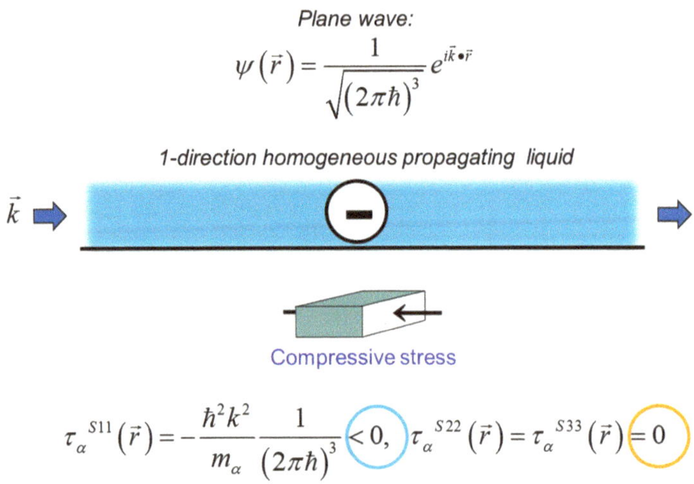

Fig. 3.10 Plane wave: one-direction homogeneous propagating liquid with the principal axis $\vec{k}/|\vec{k}|$

Stress tensor of free particle with plane wave has one negative and doubly degenerate null eigenvalues. The particle has one-direction homogeneous propagating liquid character (see Fig. 3.10).

Stress tensor for electron in the ground state under bare ionic core has degenerate negative stresses for the surface mode: two-dimensional surface liquid character with null stress in the radial mode (see Fig. 3.11).

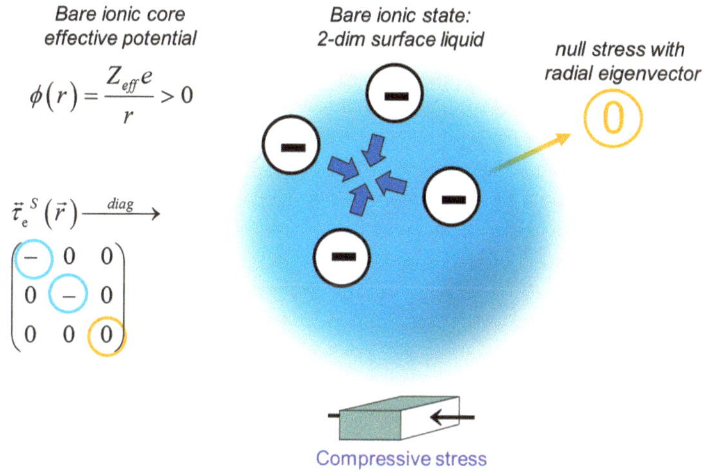

Fig. 3.11 Bare ionic state: two-dimensional surface liquid with null stress in the radial mode

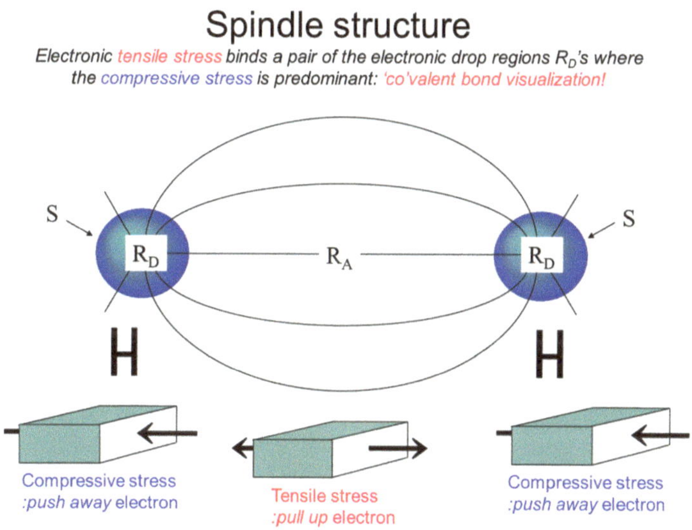

Fig. 3.12 The discovery of the "spindle structure" of the covalent bond and the long-range Lewis pair formation, a novel local picture of chemical interaction based on the electronic stress tensor

The electronic tensile stress *pulling up* electron through a surface in between is visualized as the spindle structure binding a pair of the electronic drop regions R_Ds separated from each other through the electronic atmosphere region R_A with the interface S which separates them (see Fig. 3.12).

The spindle structure is mathematically proved to appear at any region where the new Lewis electron pair is formed in association with inphase overlap of orbitals, like in between a pair of H atoms (see Fig. 3.3). The spindle structure is hidden

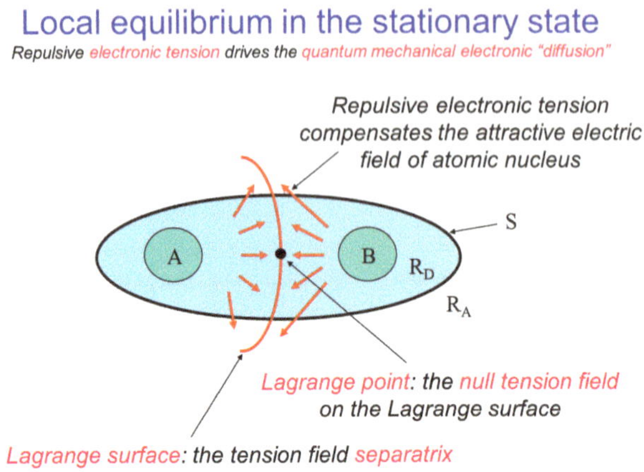

Fig. 3.13 Local equilibrium with the Lagrange surface and the Lagrange point

where out-of-phase overlap of orbitals overwhelms the former, like in between a pair of He atoms, forming the anti-spindle structure (see Fig. 3.3). The compressive stress *pushing away* electron in the remote electronic drop region R_D from the adjacent electronic atmosphere region R_A through the interface S which separates them. The consequence is the no reformation of the new Lewis pair of electron. There appears no new spindle structure. The spindle structure is also hidden where a pair of atomic nuclei is so closely combined, like in between a pair of C atoms in C_2H_2. Since the R_A, R_A, and S are measures of the kinetic energy density, which physically define the intrinsic shape volume of atoms and molecules, they are also used to define the intrinsic electronic transition state along the course of the reaction coordinate.

In a molecule AB composed of atoms A and B, the universal local equilibrium picture in the electronic stationary state is shown in Fig. 3.13. The Heisenberg uncertainty principle let electron diffuse away from each atomic center it belongs. The diffusive force is the tension $\hat{\vec{\tau}}^{S}{}_{\alpha}(\vec{r}) = \mathrm{div}\,\hat{\overleftrightarrow{\tau}}^{S}_{\alpha}(\vec{r})$ given in Eq. (3.9) compensating the Lorentz force exerting from each atomic center (see Fig. 1.30). The tension vector field collides to form separatrix which discriminates the region of atomic center. The separatrix is called the Lagrange surface; if the null tension field on the Lagrange surface, it is called the Lagrange point (see Fig. 3.13).

3.3 Stress as the Energy Density

3.3.1 *Liquid Character: Standing Wave Mode of Tensionless Electron*

Let an electron be bound in a box of rectangular cuboid with attractive potential energy (Tachibana 2014)

$$V(\vec{r}) = \begin{cases} V_{\text{inner}}(\vec{r}) < 0 \quad , & \vec{r} \in \Omega_{\text{inner}} \\ V_{\text{outer}}(\vec{r}) = 0 \quad , & \vec{r} \in \Omega_{\text{outer}} \end{cases} \tag{3.17}$$

with

$$\begin{array}{ll} \Omega_{\text{inner}}: & |x| < \ell_x, |y| < \ell_y, |z| < \ell_z \\ \Omega_{\text{outer}}: & |x| > \ell_x, |y| > \ell_y, |z| > \ell_z \end{array} \tag{3.18}$$

Let electron be in a steady state with the standing wave function

$$\psi(\vec{r}) = \begin{cases} \psi_{\text{inner}}(\vec{r}) = N_{\text{norm}}(R)\psi_x(x)\psi_y(y)\psi_z(z), & \vec{r} \in R \subset \Omega_{\text{inner}} \\ \psi_{\text{outer}}(\vec{r}) = 0, & \vec{r} \in \Omega_{\text{outer}} \end{cases} \tag{3.19}$$

$$\int_{\text{whole space}} d^3\vec{r}|\psi(\vec{r})|^2 = 1 \tag{3.20}$$

$$\psi_x(x) = \sin\left(k_x(R)(x - a_x(R))\right) \quad , \quad \vec{r} \in R \subset \Omega_{\text{inner}} \tag{3.21}$$

where we assume region-wise resolution of Ω_{inner} with real constants $N_{\text{norm}}(R)$, $k_x(R)$, and $a_x(R)$ in a region of $\vec{r} \in R \subset \Omega_{\text{inner}}$ together with similar forms for $\psi_y(y)$ and $\psi_z(z)$. Then the regional stress tensor of electron becomes block diagonal as (see Fig. 3.14)

$$\overleftrightarrow{\tau}^S_{\text{e}}(\vec{r}) = \begin{cases} \overleftrightarrow{\tau}^S_{\text{e}_{\text{inner}}}(\vec{r}), & \vec{r} \in R \subset \Omega_{\text{inner}} \\ \overleftrightarrow{\tau}^S_{\text{e}_{\text{outer}}}(\vec{r}) = 0, & \vec{r} \in \Omega_{\text{outer}} \end{cases} \tag{3.22}$$

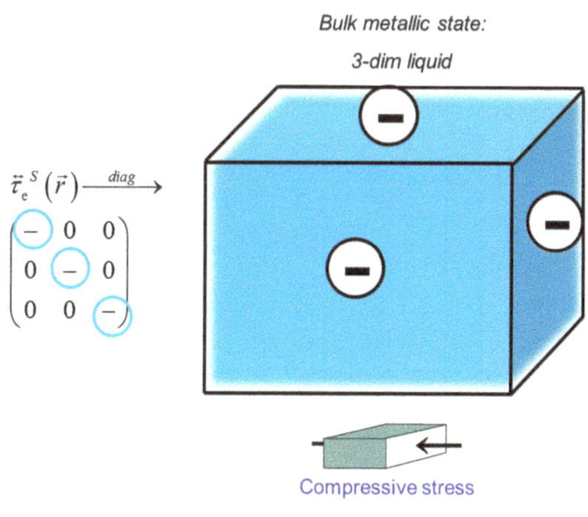

Fig. 3.14 Metallic state

$$\overleftrightarrow{\tau}^{S}_{\mathrm{e_{inner}}}(\vec{r}) = \begin{pmatrix} \tau^{Sxx}_{\mathrm{e_{inner}}}(\vec{r}) & 0 & 0 \\ 0 & \tau^{Syy}_{\mathrm{e_{inner}}}(\vec{r}) & 0 \\ 0 & 0 & \tau^{Szz}_{\mathrm{e_{inner}}}(\vec{r}) \end{pmatrix} \tag{3.23}$$

$$\begin{aligned} \tau^{Sxx}_{\mathrm{e_{inner}}}(\vec{r}) &= -\frac{\hbar^2 k_x{}^2(R)}{2m_\mathrm{e}} N_{\mathrm{norm}}{}^2(R)\left|\psi_y(y)\psi_z(z)\right|^2 < 0 \\ \tau^{Syy}_{\mathrm{e_{inner}}}(\vec{r}) &= -\frac{\hbar^2 k_y{}^2(R)}{2m_\mathrm{e}} N_{\mathrm{norm}}{}^2(R)\left|\psi_z(y)\psi_x(z)\right|^2 < 0 \\ \tau^{Szz}_{\mathrm{e_{inner}}}(\vec{r}) &= -\frac{\hbar^2 k_z{}^2(R)}{2m_\mathrm{e}} N_{\mathrm{norm}}{}^2(R)\left|\psi_x(z)\psi_y(y)\right|^2 < 0 \end{aligned} \tag{3.24}$$

Unless at the point of the node of wave function, the eigenvalues are all negative, which represents compressive stress in every direction and demonstrates the "liquid" character of the bound electron in the standing wave mode. This is compatible with the vanishment of the tension as proved to be

$$\vec{\tau}^{S}_{\mathrm{e}}(\vec{r}) = \begin{cases} \vec{\tau}^{S}_{\mathrm{e_{inner}}}(\vec{r}) = 0, & \vec{r} \in R \subset \Omega_{\mathrm{inner}} \\ \vec{\tau}^{S}_{\mathrm{e_{outer}}}(\vec{r}) = 0, & \vec{r} \in \Omega_{\mathrm{outer}} \end{cases} \tag{3.25}$$

which should be so since the internal force is null in $R \subset \Omega_{\mathrm{inner}}$ or Ω_{outer}. As to the trace of the electronic stress tensor $\overleftrightarrow{\tau}^{S}_{\mathrm{e_{inner}}}(\vec{r})$, we have

$$\frac{1}{2}\int_{\Omega_{\mathrm{inner}}} d^3\vec{r}\left(\overleftrightarrow{\tau}^{S}_{\mathrm{e_{inner}}}(\vec{r})\right)^{kk} = -\int_{\Omega_{\mathrm{inner}}} d^3\vec{r} n_{T_{\mathrm{e_{inner}}}}(\vec{r}) \tag{3.26}$$

where $n_{T_{\mathrm{e_{inner}}}}(\vec{r})$ is the kinetic energy density of electron, which is regionally found as

$$n_{T_{\mathrm{e_{inner}}}}(\vec{r}) = \frac{\hbar^2\left(k_x{}^2(R) + k_y{}^2(R) + k_z{}^2(R)\right)}{2m_\mathrm{e}}\left|\psi_{\mathrm{inner}}(\vec{r})\right|^2, \quad \vec{r} \in R \subset \Omega_{\mathrm{inner}} \tag{3.27}$$

Using the virial theorem, we confirm half the trace of $\overleftrightarrow{\tau}^{S}_{\mathrm{e}}(\vec{r})$ be the energy density of electron

$$\frac{1}{2}\int_{\text{whole space}} d^3\vec{r}\left(\overleftrightarrow{\tau}^{S}_{\mathrm{e}}(\vec{r})\right)^{kk} = E_{\mathrm{e;primary\ Rigged\ QED}} \tag{3.28}$$

with

$$E_{\mathrm{e;primary\ Rigged\ QED}} = \int_{\Omega_{\mathrm{inner}}} d^3\vec{r} n_{T_{\mathrm{e_{inner}}}}(\vec{r}) + \int_{\Omega_{\mathrm{inner}}} d^3\vec{r} V_{\mathrm{inner}}(\vec{r})\left|\psi_{\mathrm{inner}}(\vec{r})\right|^2 \tag{3.29}$$

3.3.2 *Liquid Character: Propagating Wave Mode of Tensionless Electron*

Let an electron be bound in the same box of rectangular cuboid but assume the size be extremely large

$$\ell_x, \ell_y, \ell_z >> 1 \tag{3.30}$$

so that the electron may be in the regional traveling mode with wave function

$$\psi_x(x) = \exp(ik_x(R)(x - a_x(R))) \quad , \quad \vec{r} \in R \subset \Omega_{\text{inner}} \tag{3.31}$$

with the obvious notation for $\psi_y(y)$ and $\psi_z(z)$.

Then the stress tensor of electron becomes

$$\overleftrightarrow{\tau}^S_{\text{e}}(\vec{r}) = \begin{cases} \overleftrightarrow{\tau}^S_{\text{e}_{\text{inner}}}(\vec{r}) \quad , \quad \vec{r} \in R \subset \Omega_{\text{inner}} \\ \overleftrightarrow{\tau}^S_{\text{e}_{\text{outer}}}(\vec{r}) = 0 \quad , \quad \vec{r} \in \Omega_{\text{outer}} \end{cases} \tag{3.32}$$

$$\overleftrightarrow{\tau}^S_{\text{e}_{\text{inner}}}(\vec{r}) = -\frac{\hbar^2}{m_{\text{e}}} \begin{pmatrix} k_x{}^2(R) & k_x(R)k_y(R) & k_x(R)k_z(R) \\ k_y(R)k_x(R) & k_y{}^2(R) & k_y(R)k_z(R) \\ k_z(R)k_x(R) & k_z(R)k_y(R) & k_z{}^2(R) \end{pmatrix} N_{\text{norm}}{}^2(R) \tag{3.33}$$

and the tension

$$\vec{\tau}^S_{\text{e}}(\vec{r}) = \begin{cases} \vec{\tau}^S_{\text{e}_{\text{inner}}}(\vec{r}) = 0 \quad , \quad \vec{r} \in R \subset \Omega_{\text{inner}} \\ \vec{\tau}^S_{\text{e}_{\text{outer}}}(\vec{r}) = 0 \quad , \quad \vec{r} \in \Omega_{\text{outer}} \end{cases} \tag{3.34}$$

which should be so since the internal force is null in the region of $\vec{r} \in R \subset \Omega_{\text{inner}}$ or Ω_{outer}. Diagonalizing $\vec{\tau}^S_{\text{e}_{\text{inner}}}(\vec{r})$ leads to the first negative eigenvalue and the degenerate zero eigenvalues

$$\begin{aligned} &\tau^{S11}_{\text{e}_{\text{inner}}}(\vec{r}) = -\frac{\hbar^2 \left|\vec{k}(R)\right|^2}{m_{\text{e}}} N_{\text{norm}}{}^2(R) < 0, \quad \tau^{S22}_{\text{e}_{\text{inner}}}(\vec{r}) = \tau^{S33}_{\text{e}_{\text{inner}}}(\vec{r}) = 0 \\ &\vec{k}(R) = (k_x(R), k_y(R), k_z(R)) \end{aligned} \tag{3.35}$$

with the eigenvector $\vec{k}(R)/\left|\vec{k}(R)\right|$ of the first eigenvalue corresponding to the liquid character in the propagating wave mode (see Fig. 3.9). As to the trace of the electronic stress tensor $\overleftrightarrow{\tau}^S_{\text{e}_{\text{inner}}}(\vec{r})$, we have

$$\frac{1}{2}\int_{\Omega_{\text{inner}}} d^3\vec{r}\left(\overset{\leftrightarrow}{\tau}{}^{S}_{\text{e}_{\text{inner}}}(\vec{r})\right)^{kk} = -\int_{\Omega_{\text{inner}}} d^3\vec{r} n_{T_{\text{e}_{\text{inner}}}}(\vec{r}) \tag{3.36}$$

where $n_{T_{\text{e}_{\text{inner}}}}(\vec{r})$ is the kinetic energy density of electron, which is regionally found as

$$n_{T_{\text{e}_{\text{inner}}}}(\vec{r}) = \frac{\hbar^2\left|\vec{k}(R)\right|^2}{2m_{\text{e}}}|\psi_{\text{inner}}(\vec{r})|^2 \ , \quad \vec{r}\in R\subset\Omega_{\text{inner}} \tag{3.37}$$

Using the virial theorem, we confirm half the trace of $\overset{\leftrightarrow}{\tau}{}^{S}_{\text{e}}(\vec{r})$ to be the energy density of electron

$$\frac{1}{2}\int_{\text{whole space}} d^3\vec{r}\left(\overset{\leftrightarrow}{\tau}{}^{S}_{\text{e}}(\vec{r})\right)^{kk} = E_{\text{e;Primary Rigged QED}} \tag{3.38}$$

with

$$E_{\text{e;Primary Rigged QED}} = \int_{\Omega_{\text{inner}}} d^3\vec{r} n_{T_{\text{e}_{\text{inner}}}}(\vec{r}) + \int_{\Omega_{\text{inner}}} d^3\vec{r} V_{\text{inner}}(\vec{r})|\psi_{\text{inner}}(\vec{r})|^2 \tag{3.39}$$

3.3.3 *Mixed Character: The Bloch Wave Mode of Tension Finite Electron*

Let an electron be bound in the same extremely large box of rectangular cuboid, but here we assume lattice periodicity for the potential:

$$V_{\text{inner}}(\vec{r}+\vec{T}) = V_{\text{inner}}(\vec{r}) \ , \quad \vec{r} \text{ and } \vec{r}+\vec{T}\in R\subset\Omega_{\text{inner}} \tag{3.40}$$

with $\vec{T}$ as the lattice translational vector of the molecule at each lattice point. Then, we may choose the Bloch orbital for electron (Tachibana 2014)

$$\psi(\vec{r}) = \begin{cases} \psi_{\text{inner}}(\vec{r}) = \psi_{\vec{k}}(\vec{r}) \ , & \vec{r}\in R\subset\Omega_{\text{inner}} \\ \psi_{\text{outer}}(\vec{r}) = 0 \ , & \vec{r}\in\Omega_{\text{outer}} \end{cases} \tag{3.41}$$

$$\int_{\text{whole space}} d^3\vec{r}|\psi(\vec{r})|^2 = 1 \tag{3.42}$$

$$\psi_{\vec{k}}(\vec{r}) = \exp\left(i\vec{k}(R)\bullet\vec{r}\right)u_{\vec{k}}(\vec{r}) \ , \quad u_{\vec{k}}(\vec{r}+\vec{T}) = u_{\vec{k}}(\vec{r}) \tag{3.43}$$

with the periodic function $u_{\vec{k}}(\vec{r})$.

Then the stress tensor of electron becomes periodic and decomposed into two terms each periodic:

$$\begin{aligned}
&\overleftrightarrow{\tau}^{S}_{\mathrm{e}}(\vec{r}) = \begin{cases} \overleftrightarrow{\tau}^{S}_{\mathrm{e_{inner}}}(\vec{r}) = \overleftrightarrow{\tau}^{S}_{\mathrm{e_{inner;band\text{-}like}}}(\vec{r}) + \overleftrightarrow{\tau}^{S}_{\mathrm{e_{inner;molecule\text{-}like}}}(\vec{r}), & \vec{r} \in R \subset \Omega_{\mathrm{inner}} \\ \overleftrightarrow{\tau}^{S}_{\mathrm{e_{outer}}}(\vec{r}) = 0, & \vec{r} \in \Omega_{\mathrm{outer}} \end{cases} \\
&\overleftrightarrow{\tau}^{S}_{\mathrm{e_{inner}}}(\vec{r} + \vec{T}) = \overleftrightarrow{\tau}^{S}_{\mathrm{e_{inner}}}(\vec{r})
\end{aligned} \tag{3.44}$$

where

$$\begin{aligned}
\overleftrightarrow{\tau}^{Sij}_{\mathrm{e_{inner;band\text{-}like}}}(\vec{r}) &= \frac{\hbar^2}{4m_{\mathrm{e}}} \begin{pmatrix} -2k^i(R)k^j(R)\left|u_{\vec{k}}(\vec{r})\right|^2 \\ +2ik^i(R)u_{\vec{k}}^{\dagger}(\vec{r})\dfrac{\partial}{\partial x^j}u_{\vec{k}}(\vec{r}) + 2ik^j(R)u_{\vec{k}}^{\dagger}(\vec{r})\dfrac{\partial}{\partial x^i}u_{\vec{k}}(\vec{r}) \\ +c.c. \end{pmatrix} \\
\overleftrightarrow{\tau}^{Sij}_{\mathrm{e_{inner;molecule\text{-}like}}}(\vec{r}) &= \frac{\hbar^2}{4m_{\mathrm{e}}}\left(u_{\vec{k}}^{\dagger}(\vec{r})\frac{\partial^2}{\partial x^i \partial x^j}u_{\vec{k}}(\vec{r}) - \left(\tfrac{\partial}{\partial x^i}u_{\vec{k}}(\vec{r})\right)^{\dagger}\frac{\partial}{\partial x^j}u_{\vec{k}}(\vec{r}) + c.c.\right)
\end{aligned} \tag{3.45}$$

and tension

$$\vec{\tau}^{S}_{\mathrm{e}}(\vec{r}) = \begin{cases} \mathrm{div}\,\overleftrightarrow{\tau}^{S}_{\mathrm{e_{inner}}}(\vec{r}) \text{ may not be null}, & \vec{r} \in R \subset \Omega_{\mathrm{inner}} \\ \vec{\tau}^{S}_{\mathrm{e_{outer}}}(\vec{r}) = 0, & \vec{r} \in \Omega_{\mathrm{outer}} \end{cases} \tag{3.46}$$

which should be so since the internal force may not be null in $R \subset \Omega_{\mathrm{inner}}$ but null in Ω_{outer}.

As to the trace of the electronic stress tensor $\overleftrightarrow{\tau}^{S}_{\mathrm{e_{inner}}}(\vec{r})$, we have

$$\frac{1}{2}\int_{\Omega_{\mathrm{inner}}} d^3\vec{r}\,\overleftrightarrow{\tau}^{Skk}_{\mathrm{e_{inner}}}(\vec{r}) = -\int_{\Omega_{\mathrm{inner}}} d^3\vec{r}\, n_{T_{\mathrm{e_{inner}}}}(\vec{r}) \tag{3.47}$$

where $n_{T_{\mathrm{e_{inner}}}}(\vec{r})$ is the kinetic energy density of electron, which is regionally found as

$$n_{T_{\mathrm{e_{inner}}}}(\vec{r}) = n_{T_{\mathrm{e_{inner;band\text{-}like}}}}(\vec{r}) + n_{T_{\mathrm{e_{inner;molecule\text{-}like}}}}(\vec{r}), \quad \vec{r} \in R \subset \Omega_{\mathrm{inner}} \tag{3.48}$$

$$n_{T_{e_{\text{inner;band-like}}}}(\vec{r}) = -\frac{\hbar^2}{4m_e}\left(\begin{array}{l} -2\left|\vec{k}(R)\right|^2 \left|u_{\vec{k}}(\vec{r})\right|^2 \\ +2i\vec{k}(R)\bullet\left(u_{\vec{k}}^{\dagger}(\vec{r})\vec{\nabla}u_{\vec{k}}(\vec{r}) - \left(\vec{\nabla}u_{\vec{k}}(\vec{r})\right)^{\dagger}u_{\vec{k}}(\vec{r})\right)\end{array}\right)$$

$$n_{T_{e_{\text{inner;molecule-like}}}}(\vec{r}) = -\frac{\hbar^2}{4m_e}\left(u_{\vec{k}}^{\dagger}(\vec{r})\Delta u_{\vec{k}}(\vec{r}) + \left(\Delta u_{\vec{k}}(\vec{r})\right)^{\dagger}u_{\vec{k}}(\vec{r})\right) \tag{3.49}$$

Using the virial theorem, we confirm that half the trace of $\overleftrightarrow{\tau}^{S}_{e}(\vec{r})$ be the energy density of electron

$$\frac{1}{2}\int_{\text{whole space}} d^3\vec{r}\,\overleftrightarrow{\tau}^{Skk}_{e}(\vec{r}) = E_{\text{e;primary Rigged QED}} \tag{3.50}$$

with

$$E_{\text{e;primary Rigged QED}} = \int_{\Omega_{\text{inner}}} d^3\vec{r}\, n_{T_{e_{\text{inner}}}}(\vec{r}) + \int_{\Omega_{\text{inner}}} d^3\vec{r}\, V_{\text{inner}}(\vec{r})\left|\psi_{\text{inner}}(\vec{r})\right|^2 \tag{3.51}$$

Let the behavior of $\overleftrightarrow{\tau}^{S}_{e_{\text{inner;band-like}}}(\vec{r})$ be further examined in the region far from the atomic nucleus with the negligible gradient with respect to space, $\vec{\nabla}u_{\vec{k}}(\vec{r}) \to 0$:

$$\overleftrightarrow{\tau}^{S}_{e_{\text{inner;band-like}}}(\vec{r}) \to -\frac{\hbar^2}{m_e}\begin{pmatrix} k_x{}^2(R) & k_x(R)k_y(R) & k_x(R)k_z(R) \\ k_y(R)k_x(R) & k_y{}^2(R) & k_y(R)k_z(R) \\ k_z(R)k_x(R) & k_z(R)k_y(R) & k_z{}^2(R) \end{pmatrix}\left|u_{\vec{k}}(\vec{r})\right|^2 \tag{3.52}$$

This is proportional to Eq. (3.33), demonstrating the liquid character with the propagating wave mode in the limit of constant density

$$\left|u_{\vec{k}}(\vec{r})\right|^2 \to N_{\text{norm}}{}^2(R) \tag{3.53}$$

Also let the behavior of $\overleftrightarrow{\tau}^{S}_{e_{\text{inner;molecule-likee}}}(\vec{r})$ be examined in the innermost region close to the atomic nucleus of molecule that is responsible for making $V_{\text{inner}}(\vec{r})$ periodic:

$$\overset{\leftrightarrow}{\tau}{}^{Sij}_{\mathrm{e}_{\mathrm{inner;molecule-like}}}(\vec{r}) \rightarrow \overset{\leftrightarrow}{\tau}{}^{Sij}_{\mathrm{e}_{\mathrm{molecule}}}(\vec{r})$$
$$= \frac{\hbar^2}{4m_\mathrm{e}} \left(\begin{array}{l} u_{\mathrm{molecule}}{}^{\dagger}(\vec{r}) \frac{\partial^2}{\partial x^i \partial x^j} u_{\mathrm{molecule}}(\vec{r}) - \left(\frac{\partial}{\partial x^i} u_{\mathrm{molecule}}(\vec{r})\right)^{\dagger} \frac{\partial}{\partial x^j} u_{\mathrm{molecule}}(\vec{r}) \\ +c.c. \end{array} \right) \tag{3.54}$$

Thus, the liquid character now is mixed with the molecular character: this is the stress tensor of electron adhered to the molecule, $\overset{\leftrightarrow}{\tau}{}^{S}_{\mathrm{e}_{\mathrm{molecule}}}(\vec{r})$. This $\overset{\leftrightarrow}{\tau}{}^{S}_{\mathrm{e}_{\mathrm{molecule}}}(\vec{r})$ itself may bring about the tensile stress as well as the compressive stress with non-null tension, depending on the bonding character of electron in molecule. The most prominent character is the spindle structure.

3.3.4 *Spindle Structure Along the Reaction Coordinate*

3.3.4.1 Pairing Stress

Let the $\overset{\leftrightarrow}{\tau}{}^{S}_{\mathrm{e}_{\mathrm{molecule}}}(\vec{r})$ be examined by $\overset{\leftrightarrow}{\tau}{}^{S}_{\mathrm{e}}(\vec{r})$ for simple hydrogen molecule with a pair of protons be clamped at positions $\vec{a} = \left(0, 0, -\frac{R}{2}\right)$ and $\vec{b} = \left(0, 0, \frac{R}{2}\right)$. The Lewis electron pairing is the inphase overlap of two remote electrons, the + form of the Heitler–London wave functions $u_{\pm}(\vec{r})$ for $u_{\mathrm{molecule}}(\vec{r})$ (see Fig. 3.15)

$$u_{\mathrm{molecule}}(\vec{r}) = u_{\pm}(\vec{r}) = N_{\pm}\left(e^{-\zeta r_a} \pm e^{-\zeta r_b}\right) \tag{3.55}$$

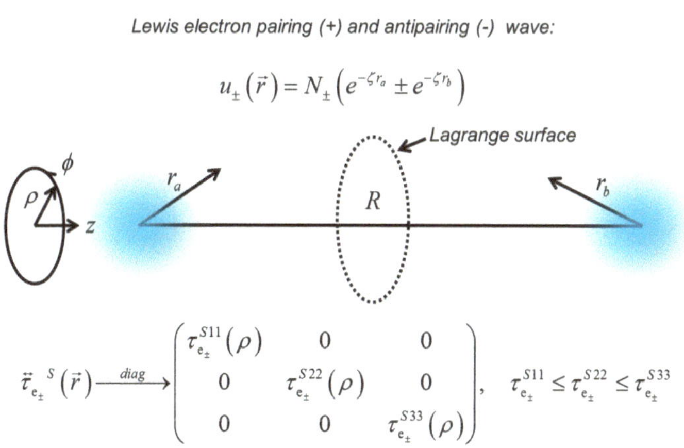

Fig. 3.15 The Lewis electron pairing (+) and antipairing (−) Heitler–London states

$$N_{\pm}(R) = \frac{d}{\sqrt{2\left(1 \pm d^2 S(R)\right)}} \tag{3.56}$$

$$d = \sqrt{\frac{1}{\int d^3\vec{r}\left(e^{-\zeta r}\right)^2}} = \sqrt{\frac{\zeta^3}{\pi}} \tag{3.57}$$

$$S(R) = \int d^3\vec{r} e^{-\zeta r_a} e^{-\zeta r_b} = \frac{\pi}{\zeta^3}\left(1 + \zeta R + \frac{1}{3}(\zeta R)^2\right) e^{-\zeta R} \tag{3.58}$$

$$R = \left|\vec{a} - \vec{b}\right| \tag{3.59}$$

$$\vec{r} = (x, y, z), \quad r = |\vec{r}| = \sqrt{x^2 + y^2 + z^2} \tag{3.60}$$

$$\vec{r}_a = \vec{r} - \vec{a} = (x_a, y_a, z_a), \quad r_a = |\vec{r}_a| = \sqrt{x^2 + y^2 + \left(z + \frac{R}{2}\right)^2} \tag{3.61}$$

$$\vec{r}_b = \vec{r} - \vec{b} = (x_b, y_b, z_b), \quad r_b = |\vec{r}_b| = \sqrt{x^2 + y^2 + \left(z - \frac{R}{2}\right)^2} \tag{3.62}$$

The kinetic energy density becomes

$$n_{T_{e\pm}}(\vec{r}) = {N_{\pm}}^2 \begin{pmatrix} -\frac{\zeta^2}{2}\left(e^{-\zeta r_a} \pm e^{-\zeta r_b}\right)^2 \\ +\zeta\left(e^{-\zeta r_a} \pm e^{-\zeta r_b}\right)\left(\frac{1}{r_a}e^{-\zeta r_a} \pm \frac{1}{r_b}e^{-\zeta r_b}\right) \end{pmatrix} \tag{3.63}$$

The stress tensor is given as

$$\tau^S{}_{e\pm xx}(\vec{r}) = \frac{{N_{\pm}}^2}{2}\left[\begin{array}{l} \left(e^{-\zeta r_a} \pm e^{-\zeta r_b}\right)\left\{\begin{array}{l} \left(-\zeta\left(\frac{1}{r_a} - \frac{{x_a}^2}{{r_a}^3}\right) + \zeta^2\frac{{x_a}^2}{{r_a}^2}\right)e^{-\zeta r_a} \\ \pm\left(-\zeta\left(\frac{1}{r_b} - \frac{{x_b}^2}{{r_b}^3}\right) + \zeta^2\frac{{x_b}^2}{{r_b}^2}\right)e^{-\zeta r_b} \end{array}\right\} \\ -\left(\zeta\frac{x_a}{r_a}e^{-\zeta r_a} \pm \zeta\frac{x_b}{r_b}e^{-\zeta r_b}\right)^2 \end{array}\right] \tag{3.64}$$

$$\tau^S{}_{e\pm xy}(\vec{r}) = \tau^S{}_{e\pm yx}(\vec{r}) = \frac{{N_\pm}^2}{2}\left[\begin{array}{l}(e^{-\zeta r_a} \pm e^{-\zeta r_b})\left\{\begin{array}{l}\left(\zeta\frac{x_a y_a}{{r_a}^3} + \zeta^2\frac{x_a y_a}{{r_a}^2}\right)e^{-\zeta r_a} \\ \pm\left(\zeta\frac{x_b y_b}{{r_b}^3} + \zeta^2\frac{x_b y_b}{{r_b}^2}\right)e^{-\zeta r_b}\end{array}\right\} \\ -\left(\zeta\frac{x_a}{r_a}e^{-\zeta r_a} \pm \zeta\frac{x_b}{r_b}e^{-\zeta r_b}\right)\left(\zeta\frac{y_a}{r_a}e^{-\zeta r_a} \pm \zeta\frac{y_b}{r_b}e^{-\zeta r_b}\right)\end{array}\right] \tag{3.65}$$

$$\tau^S{}_{e\pm xz}(\vec{r}) = \tau^S{}_{e\pm zx}(\vec{r}) = \frac{{N_\pm}^2}{2}\left[\begin{array}{l}(e^{-\zeta r_a} \pm e^{-\zeta r_b})\left\{\begin{array}{l}\left(\zeta\frac{x_a z_a}{{r_a}^3} + \zeta^2\frac{x_a z_a}{{r_a}^2}\right)e^{-\zeta r_a} \\ \pm\left(\zeta\frac{x_b z_b}{{r_b}^3} + \zeta^2\frac{x_b z_b}{{r_b}^2}\right)e^{-\zeta r_b}\end{array}\right\} \\ -\left(\zeta\frac{x_a}{r_a}e^{-\zeta r_a} \pm \zeta\frac{x_b}{r_b}e^{-\zeta r_b}\right)\left(\zeta\frac{z_a}{r_a}e^{-\zeta r_a} \pm \zeta\frac{z_b}{r_b}e^{-\zeta r_b}\right)\end{array}\right] \tag{3.66}$$

$$\tau^S{}_{e\pm yy}(\vec{r}) = \frac{{N_\pm}^2}{2}\left[\begin{array}{l}(e^{-\zeta r_a} \pm e^{-\zeta r_b})\left\{\begin{array}{l}\left(-\zeta\left(\frac{1}{r_a} - \frac{{y_a}^2}{{r_a}^3}\right) + \zeta^2\frac{{y_a}^2}{{r_a}^2}\right)e^{-\zeta r_a} \\ \pm\left(-\zeta\left(\frac{1}{r_b} - \frac{{y_b}^2}{{r_b}^3}\right) + \zeta^2\frac{{y_b}^2}{{r_b}^2}\right)e^{-\zeta r_b}\end{array}\right\} \\ -\left(\zeta\frac{y_a}{r_a}e^{-\zeta r_a} \pm \zeta\frac{y_b}{r_b}e^{-\zeta r_b}\right)^2\end{array}\right] \tag{3.67}$$

$$\tau^S{}_{e\pm yz}(\vec{r}) = \tau^S{}_{e\pm zy}(\vec{r}) = \frac{{N_\pm}^2}{2}\left[\begin{array}{l}(e^{-\zeta r_a} \pm e^{-\zeta r_b})\left\{\begin{array}{l}\left(\zeta\frac{y_a z_a}{{r_a}^3} + \zeta^2\frac{y_a z_a}{{r_a}^2}\right)e^{-\zeta r_a} \\ \pm\left(\zeta\frac{y_b z_b}{{r_b}^3} + \zeta^2\frac{y_b z_b}{{r_b}^2}\right)e^{-\zeta r_b}\end{array}\right\} \\ -\left(\zeta\frac{y_a}{r_a}e^{-\zeta r_a} \pm \zeta\frac{y_b}{r_b}e^{-\zeta r_b}\right)\left(\zeta\frac{z_a}{r_a}e^{-\zeta r_a} \pm \zeta\frac{z_b}{r_b}e^{-\zeta r_b}\right)\end{array}\right] \tag{3.68}$$

$$\tau^S{}_{e\pm zz}(\vec{r}) = \frac{{N_\pm}^2}{2}\left[\begin{array}{l}(e^{-\zeta r_a} \pm e^{-\zeta r_b})\left\{\begin{array}{l}\left(-\zeta\left(\frac{1}{r_a} - \frac{{z_a}^2}{{r_a}^3}\right) + \zeta^2\frac{{z_a}^2}{{r_a}^2}\right)e^{-\zeta r_a} \\ \pm\left(-\zeta\left(\frac{1}{r_b} - \frac{{z_b}^2}{{r_b}^3}\right) + \zeta^2\frac{{z_b}^2}{{r_b}^2}\right)e^{-\zeta r_b}\end{array}\right\} \\ -\left(\zeta\frac{z_a}{r_a}e^{-\zeta r_a} \pm \zeta\frac{z_b}{r_b}e^{-\zeta r_b}\right)^2\end{array}\right] \tag{3.69}$$

Using the stress tensor, the tension vector becomes

$$\begin{aligned}\vec{\tau}^S_{e\pm} &= \left(\tau^S{}_{e\pm x}, \tau^S{}_{e\pm y}, \tau^S{}_{e\pm z}\right)\\ &= N_\pm{}^2\zeta\left[\begin{array}{l}\left(e^{-\zeta r_a} \pm e^{-\zeta r_b}\right)\left\{\frac{1}{r_a}\left(-\frac{\zeta^2}{2}+\frac{\zeta}{r_a}+\frac{1}{r_a{}^2}\right)e^{-\zeta r_a} \pm \frac{1}{r_b}\left(-\frac{\zeta^2}{2}+\frac{\zeta}{r_b}+\frac{1}{r_b{}^2}\right)e^{-\zeta r_b}\right\}\\ -\left(\frac{1}{r_a}e^{-\zeta r_a} \pm \frac{1}{r_b}e^{-\zeta r_b}\right)\left\{\left(-\frac{\zeta^2}{2}+\frac{\zeta}{r_a}\right)e^{-\zeta r_a} \pm \left(-\frac{\zeta^2}{2}+\frac{\zeta}{r_b}\right)e^{-\zeta r_b}\right\}\end{array}\right]\\ &\quad \times (x, y, z)\end{aligned} \tag{3.70}$$

3.3.4.2 The Lagrange Surface and the Intrinsic Electronic Transition State

The Lagrange surface is then the bond bisector plane, $z = 0$, on which we use

$$x = \rho\cos\phi, \quad y = \rho\sin\phi, \quad \rho = \sqrt{x^2 + y^2} \tag{3.71}$$

and then

$$r_a = r_b = \sqrt{\rho^2 + \left(\frac{R}{2}\right)^2} \tag{3.72}$$

and the Lagrange point is the origin (see Fig. 3.13).

On the Lagrange surface, the kinetic energy density is circularly symmetric with

$$n_{T_{e+}}(\rho) = N_+{}^2\frac{4\zeta}{\sqrt{\rho^2 + \left(\frac{R}{2}\right)^2}}e^{-2\zeta\sqrt{\rho^2+\left(\frac{R}{2}\right)^2}}\left(1 - \frac{1}{2}\zeta\sqrt{\rho^2 + \left(\frac{R}{2}\right)^2}\right) \tag{3.73}$$

$$n_{T_{e-}}(\rho) = 0 \tag{3.74}$$

and the cross section of the electronic drop region R_D with the Lagrange surface for the Lewis electron pairing (+) states is a circle of radius ρ_{R_D} found as

$$\rho_{R_D}(R) = \begin{cases}\sqrt{\left(\frac{2}{\zeta}\right)^2 - \left(\frac{R}{2}\right)^2}, & R < R^\dagger\\ 0, \quad R = R^\dagger; \quad \text{null}, & R > R^\dagger\end{cases} \tag{3.75}$$

Here, the intrinsic electronic transition state is dictated by $R = R^\dagger$ with

$$R^{\dagger} = \frac{4}{\zeta} \tag{3.76}$$

for which the shape volume defined by R_D becomes disjoint for $R > R^{\dagger}$ along the reaction coordinate. If we use the standard value $\zeta = 1$, then $R^{\dagger}$ is 4 Bohr ~ 2.117 Å (see Fig. 3.7).

3.3.4.3 Stress Tensor on the Lagrange Surface

In the Lagrange surface, the stress tensor is block diagonal.

As to the + case, we have

$$\overleftrightarrow{\tau}^{S}_{e_+} = \begin{pmatrix} \tau^{S}_{e_+ xx} & \tau^{S}_{e_+ xy} & 0 \\ \tau^{S}_{e_+ yx} & \tau^{S}_{e_+ yy} & 0 \\ 0 & 0 & \tau^{S}_{e_+ zz} \end{pmatrix} \tag{3.77}$$

where

$$\tau^{S}_{e_+ xx}(\rho, \phi) = -N_+{}^2 2\zeta \frac{\rho^2 \sin^2\phi + \left(\frac{R}{2}\right)^2}{\left(\sqrt{\rho^2 + \left(\frac{R}{2}\right)^2}\right)^3} e^{-2\zeta\sqrt{\rho^2 + \left(\frac{R}{2}\right)^2}} \tag{3.78}$$

$$\tau^{S}_{e_+ xy}(\rho, \phi) = \tau^{S}_{e_+ yx}(\rho, \phi) = N_+{}^2 2\zeta \frac{\rho^2 \cos\phi \sin\phi}{\left(\sqrt{\rho^2 + \left(\frac{R}{2}\right)^2}\right)^3} e^{-2\zeta\sqrt{\rho^2 + \left(\frac{R}{2}\right)^2}} \tag{3.79}$$

$$\tau^{S}_{e_+ yy}(\rho, \phi) = -N_+{}^2 2\zeta \frac{\rho^2 \cos^2\phi + \left(\frac{R}{2}\right)^2}{\left(\sqrt{\rho^2 + \left(\frac{R}{2}\right)^2}\right)^3} e^{-2\zeta\sqrt{\rho^2 + \left(\frac{R}{2}\right)^2}} \tag{3.80}$$

$$\tau^{S}_{e_+ zz}(\rho) = N_+{}^2 2\zeta^2 \frac{\left(\frac{R}{2}\right)^2}{\rho^2 + \left(\frac{R}{2}\right)^2} \left(1 - \frac{\rho^2}{\left(\frac{R}{2}\right)^2 \zeta \sqrt{\rho^2 + \left(\frac{R}{2}\right)^2}}\right) e^{-2\zeta\sqrt{\rho^2 + \left(\frac{R}{2}\right)^2}} \tag{3.81}$$

The spindle structure is manifested as follows: $\tau^{S11}_{e_+}(\rho) < \tau^{S22}_{e_+}(\rho) < \tau^{S33}_{e_+}(\rho)$ where $\tau^{S33}_{e_+}(\rho)$ is positive within the circle of radius ρ_{spindle}, with the first mode (see Figs. 3.16, 3.17, and 3.18)

1-st stress: compressive

$$\tau_{e_+}^{S11}(\rho) = -N_+{}^2 2\zeta \frac{\rho^2 + \left(\frac{R}{2}\right)^2}{\left(\sqrt{\rho^2 + \left(\frac{R}{2}\right)^2}\right)^3} e^{-2\zeta\sqrt{\rho^2 + \left(\frac{R}{2}\right)^2}} < 0$$

Eigenvector: breathing mode $(\cos\phi, \sin\phi, 0)$

2-nd stress: compressive

$$\tau_{e_+}^{S22}(\rho) = -N_+{}^2 2\zeta \frac{\left(\frac{R}{2}\right)^2}{\left(\sqrt{\rho^2 + \left(\frac{R}{2}\right)^2}\right)^3} e^{-2\zeta\sqrt{\rho^2 + \left(\frac{R}{2}\right)^2}} < 0$$

Eigenvector: circulating mode $(-\sin\phi, \cos\phi, 0)$

Fig. 3.16 Lewis electron pairing (+) state

3-rd stress: spindle structure

$$\tau_{e_+}^{S33}(\rho) = N_+{}^2 2\zeta^2 \frac{\left(\frac{R}{2}\right)^2}{\rho^2 + \left(\frac{R}{2}\right)^2} \left(1 - \frac{\rho^2}{\left(\frac{R}{2}\right)^2 \zeta \sqrt{\rho^2 + \left(\frac{R}{2}\right)^2}}\right) e^{-2\zeta\sqrt{\rho^2 + \left(\frac{R}{2}\right)^2}}$$

$$\begin{cases} > 0 & \text{tensile} \quad \text{if} \quad \rho < \rho_{\text{spindle}} \\ < 0 & \text{compressive} \quad \text{if} \quad \rho > \rho_{\text{spindle}} \end{cases}$$

Eigenvector: spindle mode $(0, 0, 1)$

$$\rho_{\text{spindle}}(R) = \sqrt{\frac{\zeta^2\left(\frac{R}{2}\right)^4 + \sqrt{\zeta^4\left(\frac{R}{2}\right)^8 + 4\zeta^2\left(\frac{R}{2}\right)^6}}{2}}$$

$$\xrightarrow{R\to\infty} \zeta\left(\frac{R}{2}\right)^2$$

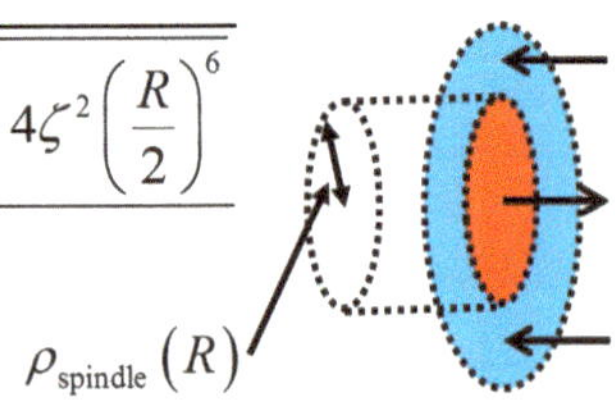

Fig. 3.17 Lewis electron pairing (+) state

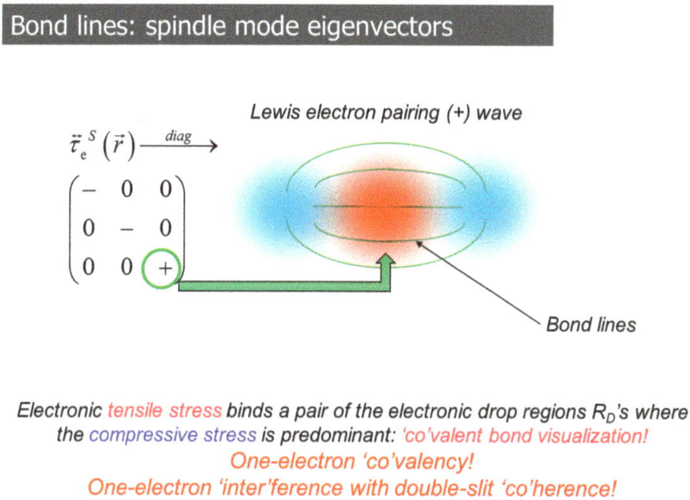

Fig. 3.18 Spindle structure

$$\tau_{e_+}^{S11}(\rho) = -N_+{}^2 2\zeta \frac{\rho^2 + \left(\frac{R}{2}\right)^2}{\left(\sqrt{\rho^2 + \left(\frac{R}{2}\right)^2}\right)^3} e^{-2\zeta\sqrt{\rho^2 + \left(\frac{R}{2}\right)^2}} < 0 \tag{3.82}$$

as the breathing mode with the principal axis $(\cos\phi, \sin\phi, 0)$, the second mode

$$\tau_{e_+}^{S22}(\rho) = -N_+{}^2 2\zeta \frac{\left(\frac{R}{2}\right)^2}{\left(\sqrt{\rho^2 + \left(\frac{R}{2}\right)^2}\right)^3} e^{-2\zeta\sqrt{\rho^2 + \left(\frac{R}{2}\right)^2}} < 0 \tag{3.83}$$

as the circulating mode with the principal axis $(-\sin\phi, \cos\phi, 0)$, and the third mode

$$\tau_{e_+}^{S33}(\rho) = \tau^{S}{}_{e_+ zz}(\rho); \quad > 0, \quad \rho < \rho_{\text{spindle}} \tag{3.84}$$

$$\rho_{\text{spindle}}(R) = \sqrt{\frac{\zeta^2\left(\frac{R}{2}\right)^4 + \sqrt{\zeta^4\left(\frac{R}{2}\right)^8 + 4\zeta^2\left(\frac{R}{2}\right)^6}}{2}} \tag{3.85}$$

as the spindle mode with the principal axis $(0, 0, 1)$.

It should be noted that the ρ_{spindle} diverges to infinity as $R \to \infty$:

$$\rho_{\text{spindle}}(R) \xrightarrow{R\to\infty} \zeta\left(\frac{R}{2}\right)^2 \tag{3.86}$$

On the other limit, ρ_{spindle} shrinks to ρ_{R_D} in Eq. (3.75) from above at $R = R_c$ that is smaller than $R = R^\dagger$ with

$$R_c = \frac{4}{\sqrt{3}\zeta} = \frac{1}{\sqrt{3}} R^\dagger < R^\dagger \tag{3.87}$$

and further to zero as $R \to 0$:

$$\rho_{\text{spindle}}(R) \xrightarrow{R \to R_c} \rho_{\text{R}_\text{D}}(R_c); \xrightarrow{R \to 0} 0 \tag{3.88}$$

As to the − case, we have

$$\overleftrightarrow{\tau}^S_{\text{e}_-} = \begin{pmatrix} 0 & 0 & 0 \\ 0 & 0 & 0 \\ 0 & 0 & \tau^S{}_{\text{e}_- zz} \end{pmatrix} \tag{3.89}$$

where

$$\tau^S{}_{\text{e}_- zz}(\rho) = -N_-{}^2 2\zeta^2 \frac{\left(\frac{R}{2}\right)^2}{\rho^2 + \left(\frac{R}{2}\right)^2} e^{-2\zeta\sqrt{\rho^2 + \left(\frac{R}{2}\right)^2}} < 0 \tag{3.90}$$

We have the anti-spindle structure as (see Fig. 3.19)

$$\tau^{S11}_{\text{e}_-}(\rho) = \tau^S{}_{\text{e}_- zz}(\rho) < 0 \tag{3.91}$$

$$\tau^{S22}_{\text{e}_-}(\rho) = \tau^{S33}_{\text{e}_-}(\rho) = 0 \tag{3.92}$$

3.3.4.4 Stress Tensor Along the Bond Axis: $x = y = 0$

The stress tensor along the bond axis $x = y = 0$ is symmetric with respect to z and block diagonal.

As to the + case, we have

$$\overleftrightarrow{\tau}^S_{\text{e}_+} = \begin{pmatrix} \tau^S{}_{\text{e}_+ xx} & 0 & 0 \\ 0 & \tau^S{}_{\text{e}_+ yy} & 0 \\ 0 & 0 & \tau^S{}_{\text{e}_+ zz} \end{pmatrix} \tag{3.93}$$

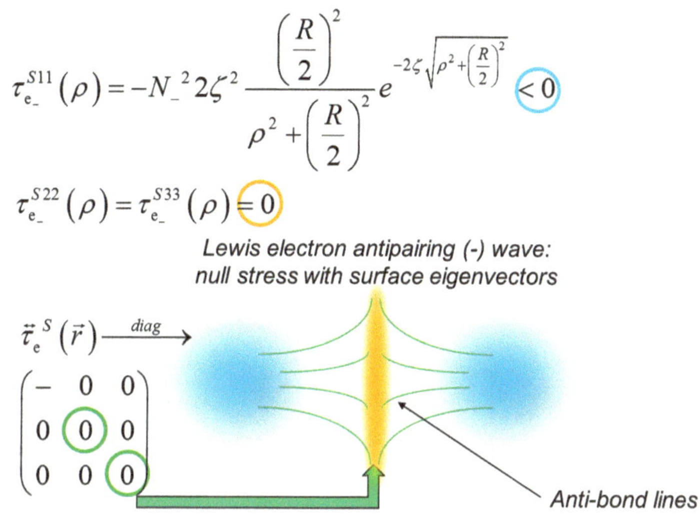

Fig. 3.19 The Lewis electron antipairing (−) state with the anti-spindle structure

$$\begin{aligned}
{\tau^S}_{e_+xx}(z) &= {\tau^S}_{e_+yy}(z) \\
&= -\frac{{N_+}^2}{2}\zeta\left(e^{-\zeta\left|z+\frac{R}{2}\right|} + e^{-\zeta\left|z-\frac{R}{2}\right|}\right)\left(\frac{1}{\left|z+\frac{R}{2}\right|}e^{-\zeta\left|z+\frac{R}{2}\right|} + \frac{1}{\left|z-\frac{R}{2}\right|}e^{-\zeta\left|z-\frac{R}{2}\right|}\right)
\end{aligned} \tag{3.94}$$

$${\tau^S}_{e_+zz}(z) = \frac{{N_+}^2}{2}\zeta^2\left\{\begin{array}{l} \left(e^{-\zeta\left|z+\frac{R}{2}\right|} + e^{-\zeta\left|z-\frac{R}{2}\right|}\right)^2 \\ -\left(\frac{z+\frac{R}{2}}{\left|z+\frac{R}{2}\right|}e^{-\zeta\left|z+\frac{R}{2}\right|} + \frac{z-\frac{R}{2}}{\left|z-\frac{R}{2}\right|}e^{-\zeta\left|z-\frac{R}{2}\right|}\right)^2 \end{array}\right\} \tag{3.95}$$

The spindle structure is manifested as

$$\tau_{e_+}^{S11}(z) = \tau_{e_+}^{S22}(z) = {\tau^S}_{e_+xx}(z) = {\tau^S}_{e_+yy}(z) < 0 \tag{3.96}$$

$$\begin{aligned}
\tau_{e_+}^{S33}(z) &= {\tau^S}_{e_+zz}(z) \\
&= \begin{cases} 2{N_+}^2\zeta^2 e^{-\zeta R} > 0, & -\frac{R}{2} < z < \frac{R}{2} \\ 0, \quad z < -\frac{R}{2} \text{ or } \frac{R}{2} < z & \end{cases}
\end{aligned} \tag{3.97}$$

Note the constancy $2{N_+}^2\zeta^2 e^{-\zeta R}$ of the eigenvalue of the spindle mode within the bond axis $-\frac{R}{2} < z < \frac{R}{2}$.

As to the − case, we have

$$\overleftrightarrow{\tau}^S_{\mathrm{e}_-} = \begin{pmatrix} \tau^S{}_{\mathrm{e}_- xx} & 0 & 0 \\ 0 & \tau^S{}_{\mathrm{e}_- yy} & 0 \\ 0 & 0 & \tau^S{}_{\mathrm{e}_- zz} \end{pmatrix} \tag{3.98}$$

$$\begin{aligned} \tau^S{}_{\mathrm{e}_- xx}(z) &= \tau^S{}_{\mathrm{e}_- yy}(z) \\ &= -\frac{N_-{}^2}{2}\zeta\left(e^{-\zeta\left|z+\frac{R}{2}\right|} - e^{-\zeta\left|z-\frac{R}{2}\right|}\right)\left(\frac{1}{\left|z+\frac{R}{2}\right|}e^{-\zeta\left|z+\frac{R}{2}\right|} - \frac{1}{\left|z-\frac{R}{2}\right|}e^{-\zeta\left|z-\frac{R}{2}\right|}\right) \end{aligned} \tag{3.99}$$

$$\tau^S{}_{\mathrm{e}_- zz}(z) = \frac{N_-{}^2}{2}\zeta^2\left\{\begin{aligned} &\left(e^{-\zeta\left|z+\frac{R}{2}\right|} - e^{-\zeta\left|z-\frac{R}{2}\right|}\right)^2 \\ &-\left(\frac{z+\frac{R}{2}}{\left|z+\frac{R}{2}\right|}e^{-\zeta\left|z+\frac{R}{2}\right|} - \frac{z-\frac{R}{2}}{\left|z-\frac{R}{2}\right|}e^{-\zeta\left|z-\frac{R}{2}\right|}\right)^2 \end{aligned}\right\} \tag{3.100}$$

This is the anti-spindle structure since (see Fig. 3.19)

$$\tau^S{}_{\mathrm{e}_- xx}(z) = \tau^S{}_{\mathrm{e}_- yy}(z)\begin{cases} < 0, & z \neq 0 \\ 0, & z = 0 \end{cases} \tag{3.101}$$

$$\begin{aligned} &\tau^S{}_{\mathrm{e}_- zz}(z) \\ &= \begin{cases} -2N_-{}^2\zeta^2 e^{-\zeta R} < 0, & -\frac{R}{2} < z < \frac{R}{2} \\ 0, \quad z < -\frac{R}{2} \text{ or } \frac{R}{2} < z \end{cases} \end{aligned} \tag{3.102}$$

3.3.4.5 Stress Tensor in the United Atom Limit: $R \to 0$

Only for the + case, the exact solution in the united atom limit $R \to 0$ makes sense. Taking the limit $R \to 0$ of the stress tensor in the Lagrange surface, we have

$$\tau^S{}_{\mathrm{e}_+ xx}(\rho, \phi) = -N_+{}^2 2\zeta\frac{\sin^2\phi}{\rho}e^{-2\zeta\rho} \tag{3.103}$$

$$\tau^S{}_{\mathrm{e}_+ xy}(\rho, \phi) = \tau^S{}_{\mathrm{e}_+ yx}(\rho, \phi) = N_+{}^2 2\zeta\frac{\cos\phi\sin\phi}{\rho}e^{-2\zeta\rho} \tag{3.104}$$

$$\tau^S{}_{\mathrm{e}_+ yy}(\rho, \phi) = -N_+{}^2 2\zeta\frac{\cos^2\phi}{\rho}e^{-2\zeta\rho} \tag{3.105}$$

$$\tau^S{}_{\mathrm{e}_+ zz}(\rho) = -N_+{}^2 2\zeta\frac{1}{\rho}e^{-2\zeta\rho} \tag{3.106}$$

In this limit, the spindle structure disappears

$$\tau_{e_+}^{S11}(\rho) = \tau_{e_+}^{S22}(\rho) = -N_+{}^2 2\zeta \frac{1}{\rho} e^{-2\zeta\rho} < 0 \tag{3.107}$$

of the degenerate surface-circulating modes with the principal axes $(-\sin\phi, \cos\phi, 0)$ and $(0, 0, 1)$, and

$$\tau_{e_+}^{S33}(\rho) = 0 \tag{3.108}$$

of the spherical-breathing mode with the principal axis $(\cos\phi, \sin\phi, 0)$. This is again for the surface liquid with $Z_{\text{eff}} = \frac{\zeta\hbar^2}{m_e e^2}$ (see Fig. 3.11).

3.3.5 *The Generic Lewis Pair Formation and the Nonclassical Bond Order*

The concept of covalency has long been known as a central guide in understanding chemical bond. The term "chemical bond" describes chemical force to realize the structure of compounds by specifying configuration of atoms in molecules. The chemical force has conventionally been described by the "attractive" force that drives atomic nuclei nearby with each other. The attractive force is the origin of chemical reactivity of atoms and molecules over and above the electrostatic repulsive force of atomic nuclei. The attractive force acting on a pair of reactants is synonymous to the chemical reactivity of the covalent bond formation.

The local stress of QED gives a unified scheme for generalized chemical reactivity. The chemical reactivity in this scheme is the force acting on a pair of electronic drop regions of reactants. This is a new kind of force acting on electrons not on nuclei. A new look at the chemical bond is elucidated where covalency is the rule of the new Lewis electron pair formation therein.

In QED the field carries the force. We are interested in the region where the force is tensile, attracting the neighboring region through the interface. If the force is repulsive through the interface, then the force is compressive. The discrimination of the nature of the field is crucial for understanding the covalent bond. The covalent bond is the region where the field is tensile, attracting the neighboring region through the interface. On the contrary, if the field is compressive, then in the region, the electrons are repulsive through the interface. Here we analyze the principal stress. If the principal stress is positive, then it is called "tensile"; if it is negative, it is called "compressive." The covalent bonding is brought about by the tensile principal stress. The force exerted on electron in the spindle structure dictates the covalent $1s\sigma$ bond. On the other hand, the antibonding orbital interaction $1s\sigma^*$results in the repulsive orbital interaction (see Fig. 3.20).

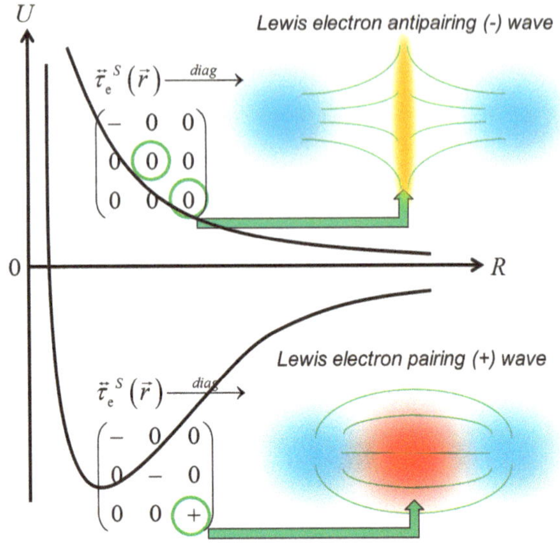

Fig. 3.20 Potential energy curve U with the $1s\sigma$ bonding and $1s\sigma^*$antibonding orbital interaction

Furthermore, using the hydrogen-like atomic orbitals for the Heitler–London wave function, it has been proved analytically that the bonding orbital interaction, such as $p\sigma$ or $p\pi$ type, does exhibit spindle structure that bridges disjoint electronic drop regions, whereas the antibonding orbital interaction, such as $p\sigma^*$ or $p\pi^*$ type, does not. Thus, the spindle structure gives universal image of the covalent bonding.

Unlike covalency, metallicity may not be measured by a single bond order. In case of metallic interaction, imagine a pair of metallic atoms with unpaired electron situated far away from each other. The tensile stress pulls up electron in the remote electronic drop region R_D to the adjacent electronic atmosphere region R_A through the interface S which separates them. The consequence is the formation of the new Lewis pair of electron. The spindle structure of covalency is universal even in this sense. However, this fact demonstrates that metal atom itself may not be sufficient to determine metallicity, where the question itself may be even meaningless. This is because metallicity may be the property of the condensed matter. Actually, as the distant pair of metallic atoms comes closer, metallicity of the condensed matter is the rule of unbinding the Lewis electron pair ever once formed. Finally, in the condensed matter, we may not be able to observe the spindle structure of covalency in its strict sense. But if an atom is going to be separated from bulk metal, then the spindle structure should emerge. This proves the emergence of covalency prerequisite to condense separated atoms into the bulk metal. The metallicity may be characterized by liquid with isotropic compressive stress in the ultimate case. The electrons contributing to the bulk metallicity behave like gluon that binds quarks in such a way that in metal bulk as condensed matter, the bond order may be small and behave as weak bond, but once if an atom is going to be separated from bulk metal, the spindle structure appears as if the bond should behave to be very strong. In other words, the bulk metallicity emerges as the long-range intrinsic electronic transition state associated with the spindle structure: the long-range Lewis pair formation

For chemical reaction is the mixture of the Lewis electron pairing and antipairing, ionicity, metallicity, lone pair, exchange repulsion, and inert gas interaction.

The regional energy decomposition is then found to be the integration over region R of space as follows:

$$E = \sum_{\mathrm{R}} E_{\mathrm{R}}, \quad E_{\mathrm{R}} = \int_{\vec{r} \in \mathrm{R}} d^3\vec{r}\,\varepsilon_\tau^S(\vec{r}) \tag{3.109}$$

$$\varepsilon_\tau^S(\vec{r}) = \frac{1}{2}\sum_\alpha \left\langle \hat{\overleftrightarrow{\tau}}_{\alpha;\text{Primary Rigged QED}}^{Skk}(\vec{r}) \right\rangle = \frac{1}{2}\sum_\alpha \tau_\alpha{}^{Skk}(\vec{r}) \tag{3.110}$$

where $\vec{r} \in \mathrm{R}$ means the regional integral confined within the region R exclusively. If the whole space is decomposed into a set of regions, where $\sum_{\mathrm{R}}$ denotes summation over the set of regions.

Likewise, the electron number N is defined and decomposed regionally as follows:

$$N = \sum_{\mathrm{R}} N_{\mathrm{R}}, \quad N_{\mathrm{R}} = \int_{\vec{r} \in \mathrm{R}} d^3\vec{r}\,n(\vec{r}) \tag{3.111}$$

$$n(\vec{r}) = \sum_\alpha \left\langle \hat{N}_{\alpha;\text{primary Rigged QED}}(\vec{r}) \right\rangle \tag{3.112}$$

As shown in Fig. 3.21, the regional energy decomposition gives the nonclassical bond order. The integral of the sum is compared with the Born–Oppenheimer electronic energy (see Fig. 3.21).

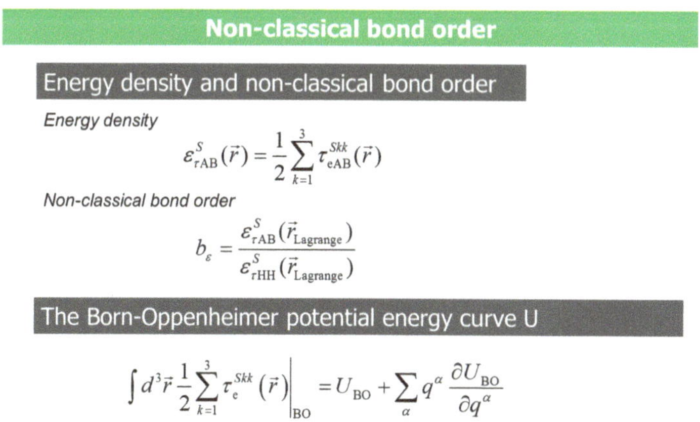

Fig. 3.21 Regional energy density and nonclassical bond order

It should be noted again that this does not mean that we are working on the Born–Oppenheimer adiabatic approximation of quantum mechanics. As shown in Eq. (3.110), the integral on the trace of nuclear stress tensor, $\int d^3\vec{r}\frac{1}{2}\sum_{a}\sum_{k=1}^{3}\tau_a{}^{Skk}(\vec{r})$, should of course be added to the energetics in the whole space.

3.4 Regional Chemical Potential

The regional chemical potential inequality principle (Tachibana 1999a) refers to (see Fig. 3.22):

- The formalization of the nonequilibrium thermodynamics based on the Onsagar local equilibrium hypothesis (Onsager 1931a, b) with considering quantum mechanical interaction through the interface between separated regions.
- The formalization of the relationship among the regional electric chemical potentials in the chemical equilibrium system by defining the regional electron numbers, the regional electronic energies, and new properties about quantum mechanical interference effect between separated regions ("the quantum mechanical law of mass action").
- The regional chemical potentials are not necessarily equal to each other, to the Fermi level of standing wave on the whole system, nor to the chemical potential of the whole system even in the chemical equilibrium system ("the regional chemical potential inequality principle").

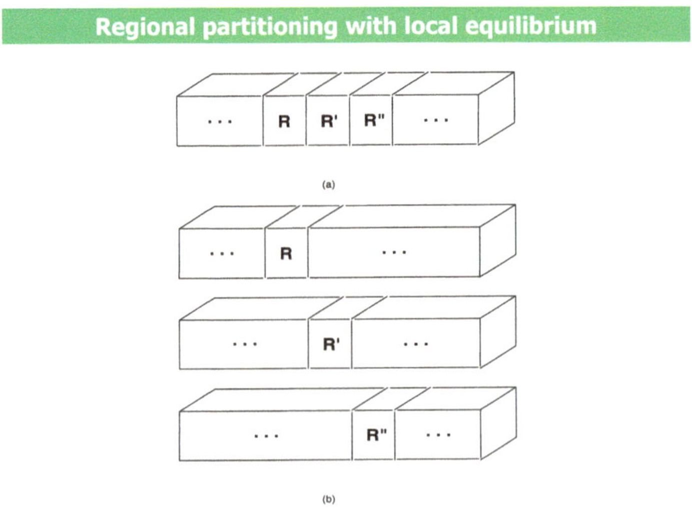

Fig. 3.22 Regional partitioning with the Onsager local equilibrium: (**a**) interface in the system A and (**b**) the Onsager hypothesis of local equilibrium

Let us consider a chemical reaction system A embedded in a medium M, an environmental system of chemical reaction (see Fig. 1.26). During the progress of chemical reaction in A, the electronic subsystem of A can exchange heat, work, and electrons with M through an interface which divides A from M. Moreover, the electronic subsystem of A can exert work on the nuclear subsystem of A and vice versa, where the nuclear subsystem of A is assumed to be thermally isolated from the medium M as well as from the electronic subsystem of A. This is the adiabatic approximation that we rely on throughout this discussion, and we neglect relativistic effects as well. If the electronic subsystem of A is in chemical equilibrium with the medium M and the chemical reaction in A is a quasi-static process, then the maximum work is gained from the electronic subsystem of A, and therefore only the minimum work is required for the nuclear subsystem of A. Gibbs proved the usefulness of the constant chemical potential between two regions in space where we observe no flux of particles whatsoever when chemical equilibrium is attained globally (Landau and Lifshitz 1980). The constancy of the chemical potential is perturbed if we put an object between a pair of regions, when the transfer of particles is rather inhibited through the interface, bringing about a finite difference in regional chemical potentials even after chemical equilibrium is attained globally (Reichl 1980).

It is not, however, a trivial matter to "observe" the inhomogeneity of the regional chemical potentials using appropriate apparatus. A promising candidate for this kind of measurement may be found in a study of the work function of metals as a function of crystallographic planes (Wigner and Bardeen 1935). The medium M in this measurement of the work function is used to observe the electrostatic potential energy of an electron at a point in the neighborhood of the crystal surface plane just outside of it (Bardeen 1936), where a clever choice of apparatus could allow the chemical potential inequality principle to be proved. However, we are not in a position here in this article to invent a device if the medium M is situated in such a way as to discriminate against the regional chemical potentials. Rather, we shall devise a method to probe the transfer of electrons within the electronic subsystem of A from one region to another through the interface situated in-between. The subdivision of the electronic subsystem of A into regions R, R′, R″, and so on is shown schematically in Fig. 3.22. In a region, R say, the electronic subsystem of A is assumed to be in chemical equilibrium, but we allow irreversible electron flow through the interface that divides R and the adjacent region, R′ say. This situation is nothing but the local equilibrium hypothesis due to Onsager (1931a, b) and is adapted in this article in order to treat irreversible electron transfer in the electronic subsystem of A (see Fig. 3.22).

We are in a position to apply the electronic tensile stress analysis to the present problem.

For example, removal of electron from the system to the reservoir gives the Gibbs chemical potential μ_G. Using Gibbs grand canonical ensemble, we arrive at the expression of the Gibbs chemical potential μ_G as follows (see Fig. 3.23):

$$\mu_G = \mu_R + \sum_{R'(\neq R)} \alpha_{R'R} \tag{3.113}$$

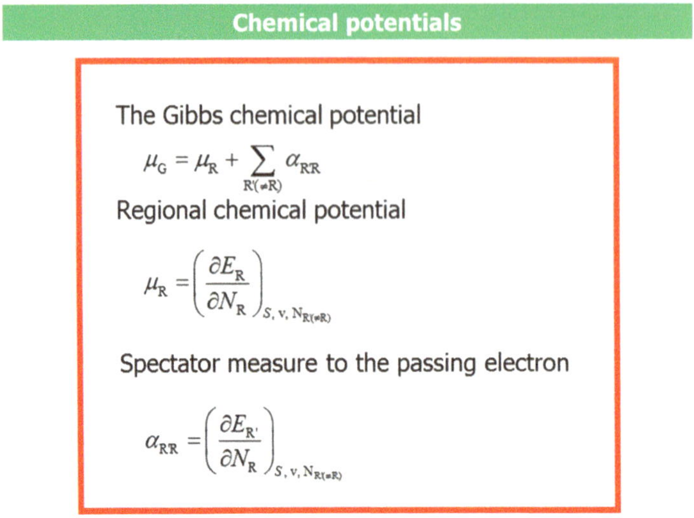

Fig. 3.23 The Gibbs chemical potential and the regional chemical potential

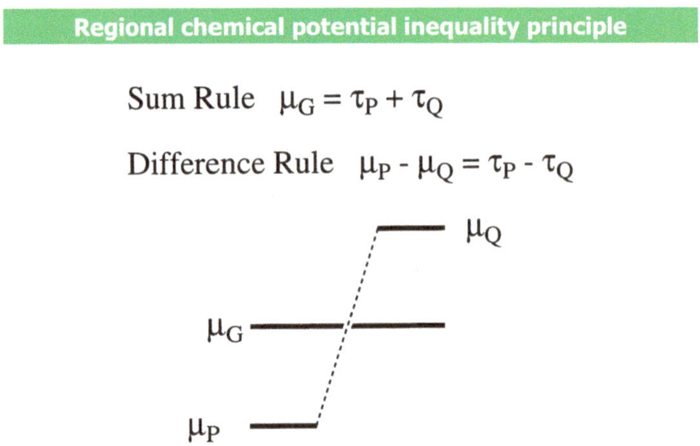

Fig. 3.24 Regional chemical potential inequality principle for two-region case

where

$$\mu_R = \left(\frac{\partial E_R}{\partial N_R}\right)_{S,v,N_{R'(\neq R)}} \tag{3.114}$$

$$\alpha_{R'R} = \left(\frac{\partial E_{R'}}{\partial N_R}\right)_{S,v,N_{R'(\neq R)}} \tag{3.115}$$

The two-region case is found in Fig. 3.24.

The regional chemical potential μ_{R} refers to the regional contribution to the μ_{G}. If an electron is withdrawn from a region R and reach the reservoir, the regional electronic energy E_{R} changes and the μ_{R} gives the energy change per one electron. On the other hand, the passage of the electron through the system to the reservoir should inevitably influence the electronic energies of the other regions, Rs, where the regional electronic energy $E_{\mathrm{R}'}$ changes and the $\alpha_{\mathrm{R}'\mathrm{R}}$ gives the energy change per one electron.

If an electron is withdrawn from a region R and passes through another region R′, then the electron is treated here external to the region R′. Namely, the R′ here acts as a spectator to the passing electron and therefore the energy change in $E_{\mathrm{R}'}$ as measured by $\alpha_{\mathrm{R}'\mathrm{R}}$ should then be "electrostatic" in nature. This should be electric potential first observed by Volta as proved by Herring and Nichols. It should be noted that manipulation of electron solely in the particular region leads to the thermodynamic definition of work function. As demonstrated by Volta for a pair of regions R and R′ in contact with each other, the contact potential difference is the difference in the regional work function as proved by Herring and Nichols (1949)

$$\phi_{\mathrm{R}} - \phi_{\mathrm{R}'} = \Phi_{\mathrm{R}'} - \Phi_{\mathrm{R}} \tag{3.116}$$

where ϕ_{R} denotes the Volta electric potential for the region R and Φ_{R} denotes the work function of the region R. This is the consequence of the chemical equilibrium in between a pair of regions in contact with each other

$$\mu_{\mathrm{G}} = -e\Phi_{\mathrm{R}} - e\phi_{\mathrm{R}} = -e\Phi_{\mathrm{R}'} - e\phi_{\mathrm{R}'} \tag{3.117}$$

where the Gibbs chemical potential μ_{G} is constant from region to region in contact with each other under the condition of global chemical equilibrium (see Fig. 3.25).

Regional chemical potential and work function

The Volta electric potential difference

$$\varphi_R - \varphi_{R'} = \Phi_{R'} - \Phi_R$$

φ_{R} : the intrinsic Volta electric potential

Φ_{R} : the intrinsic Herring-Nichols work function

The Gibbs chemical potential

$$\mu_{\mathrm{G}} = -e\Phi_{\mathrm{R}} - e\varphi_{\mathrm{R}} = -e\Phi_{\mathrm{R}'} - e\varphi_{\mathrm{R}'}$$

$$-e\varphi_{\mathrm{R}} = \sum_{\mathrm{R}'(\neq \mathrm{R})} \alpha_{\mathrm{RR}}$$

$$-e\Phi_{\mathrm{R}} = \mu_{\mathrm{R}}$$

Fig. 3.25 The Volta electric potential and the Herring and Nichols work function

Another non-classical bond order with chemical potential

Energy density per electron and another non-classical bond order

Energy density per electron

$$\left(\frac{\varepsilon_{\tau}^{S}(\vec{r})}{n(\vec{r})}\right)_{\vec{r}\in \mathrm{R}} \approx \mu_{\mathrm{R}}(\vec{r})_{\vec{r}\in \mathrm{R}}$$

Another non-classical bond order

$$b_{\mu} \approx \frac{\varepsilon_{\tau\mathrm{AB}}^{S}(\vec{r}_{\mathrm{Lagrange}})/n_{\mathrm{AB}}(\vec{r}_{\mathrm{Lagrange}})}{\varepsilon_{\tau\mathrm{HH}}^{S}(\vec{r}_{\mathrm{Lagrange}})/n_{\mathrm{HH}}(\vec{r}_{\mathrm{Lagrange}})} \approx \frac{\mu_{\mathrm{AB}}(\vec{r}_{\mathrm{Lagrange}})}{\mu_{\mathrm{HH}}(\vec{r}_{\mathrm{Lagrange}})}$$

Fig. 3.26 Another nonclassical bond order based on the regional chemical potential

On the other hand, the electrostatic effect is long-ranged. As a matter of fact, the electron is negatively charged and has non-negligible interaction with even for spectator region R″ not directly in contact with the region R. For example, the surface dipole of the spectator region R′ can contribute to the long-ranged electrostatic interaction with the electron. Hence, in our theory, the intrinsic Volta electric potential ϕ_{R} for the region R is expressed by the sum of $\alpha_{\mathrm{R'R}}$ over the spectator complementary regions R′ to R:

$$-e\phi_{\mathrm{R}} = \sum_{\mathrm{R}'(\neq \mathrm{R})} \alpha_{\mathrm{R'R}} \tag{3.118}$$

Therefore, we arrive at the intrinsic Herring–Nichols work function Φ_{R} for the region R as follows:

$$-e\Phi_{\mathrm{R}} = \mu_{\mathrm{R}} \tag{3.119}$$

Thus, the thermodynamic extension of the electronic energy density $n_E(\vec{r})$ turned out to be observable in electrochemistry: in terms of the intrinsic Volta electric potential φ_{R} and the intrinsic Herring–Nichols work function Φ_{R}. Even for the same crystal, the φ_{R} and the Φ_{R} are dependent on surface morphologies or crystallographic orientations, while the sum of these gives the constant value μ_{G} for the same crystal.

The regional chemical potential inequality principle gives another nonclassical bond order of chemical reactivity (see Fig. 3.26).

3.5 Examples

3.5.1 Chemical Bond

Shape volume is studied with the static dielectric properties of high-k dielectric materials (Nakamura et al. 2005). Numerical examples are cluster models of silicon dioxide and silicate oxides (see Fig. 3.27).

The constituent atomic shape volumes are studied from the viewpoint of the shell structure (Nozaki et al. 2016). Numerical data are atomic numbers 1 through 18 (see Fig. 3.28).

Spindle structure of chemical reaction is studied with the shape volume along the intrinsic reaction coordinate (IRC) (Tachibana 2001). Numerical example is C_2H_2+ HF (see Fig. 3.29).

Marginal stability around atoms is represented by compressive principal stresses. Only the bonds of a pair of electrons are singled out as the spindle structure, like for C–C bond, C–H bond, and H–F bond. Apparently, the tensile

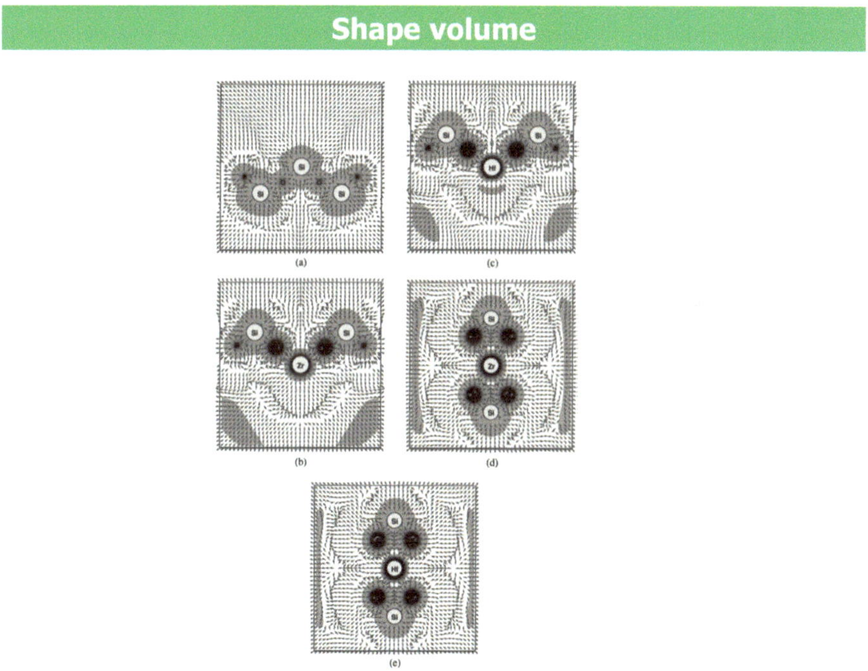

Fig. 3.27 Shape volumes of cluster models containing (**a**) three Si atoms, (**b**) Zr atom and two Si atoms, (**c**) Hf atom and two Si atoms, (**d**) Zr atom and two Si atoms in siladioxyl groups, and (**e**) Hf atom and two Si atoms in siladioxyl groups

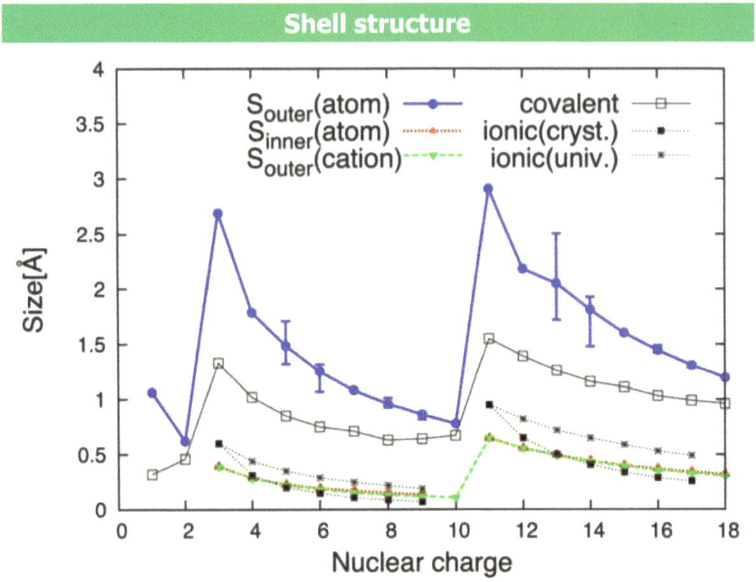

Fig. 3.28 The size of the electronic interfaces, which is defined in section "Size of the electronic interface and comparison with atomic and ionic radii," are plotted for S_{outer} (*blue solid line*) and S_{inner} (*red dotted line*) of atoms, and S_{outer} of cations (*green dashed line*). Covalent bond radii from Pyykkö and Atsumi (2009) (*black thin solid line*) and ionic radii from Pauling (1960) (*black thin dotted line*) are plotted, too. As for the ionic radii, the crystal radii (*line with filled square*) and univalent radii (*line with asterisk*) are plotted

stress regions for the C–C σ and π bonds in C_2H_2 are immersed completely under the atomic compressive ones. This is because the π bonds that spread perpendicular to the molecular plane makes the C–C distance shorter, and therefore the C–C σ bond approaches the united atom limit where the tensile stress region is immersed under the atomic compressive one. However, it should be noted that the spindle structure for the C–C σ bond is recovered in the CH_2CHF case instead. This is because one of the π bonds in C_2H_2 is lost with the reaction with HF, making the C–C distance longer (Tachibana 2005).

The π bond is manifest as the dumbbell-type spindle structure, called a sheath structure (Tachibana 2005). As in C_2H_4, the dumbbell-type spindle structure for the C–C π bond spreads in a wider region than that of the C–C σ bond and shows the magnitude of the tensile eigenvalue has two maxima in the symmetric position out of the C–C axis (see Fig. 3.30). Since the spindle structure could be a one-electron orbital property, it should be immersed in C_2H_2 (see Fig. 3.30) or found intact in H_2^+ (Ichikawa and Tachibana 2009).

The immersed spindle structure may be called the pseudo-spindle structure, while the non-closed spindle structure may be called the pro-spindle structure (Szarek et al. 2009).

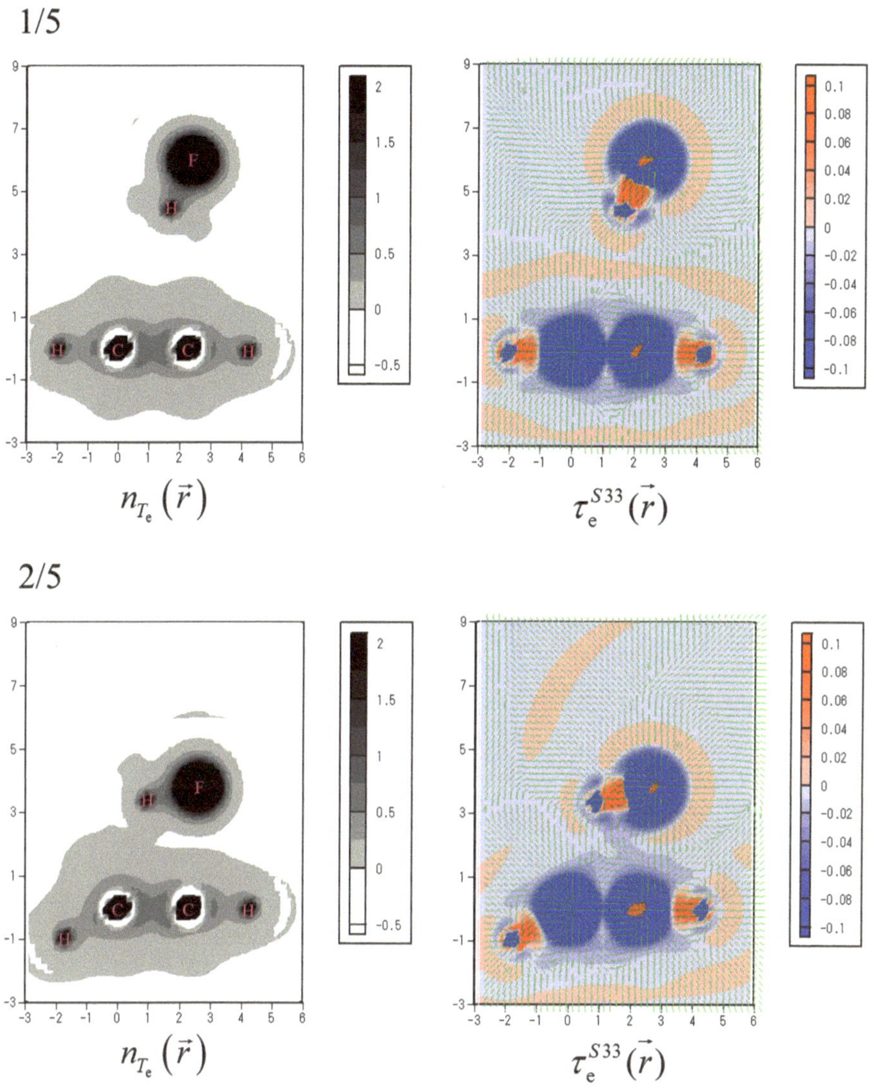

Fig. 3.29 Spindle structure with the shape volume along the IRC of HF + $C_2H_2 \rightarrow CH_2 = CHF$ in five panels. In each panel are shown the kinetic energy density (*left*), the third principal stress, and the third principal axis (*right*). The third panel corresponds to the transition state of the nuclear motion

The homonuclear diatomic molecules in ground states of main group elements, from first to fourth period, have been analyzed here with respect to the Lagrange point (Szarek and Tachibana 2007). Among the elements of the first two groups, one can notice that from Na_2 to Ca_2, the degeneracy of the largest eigenvalue

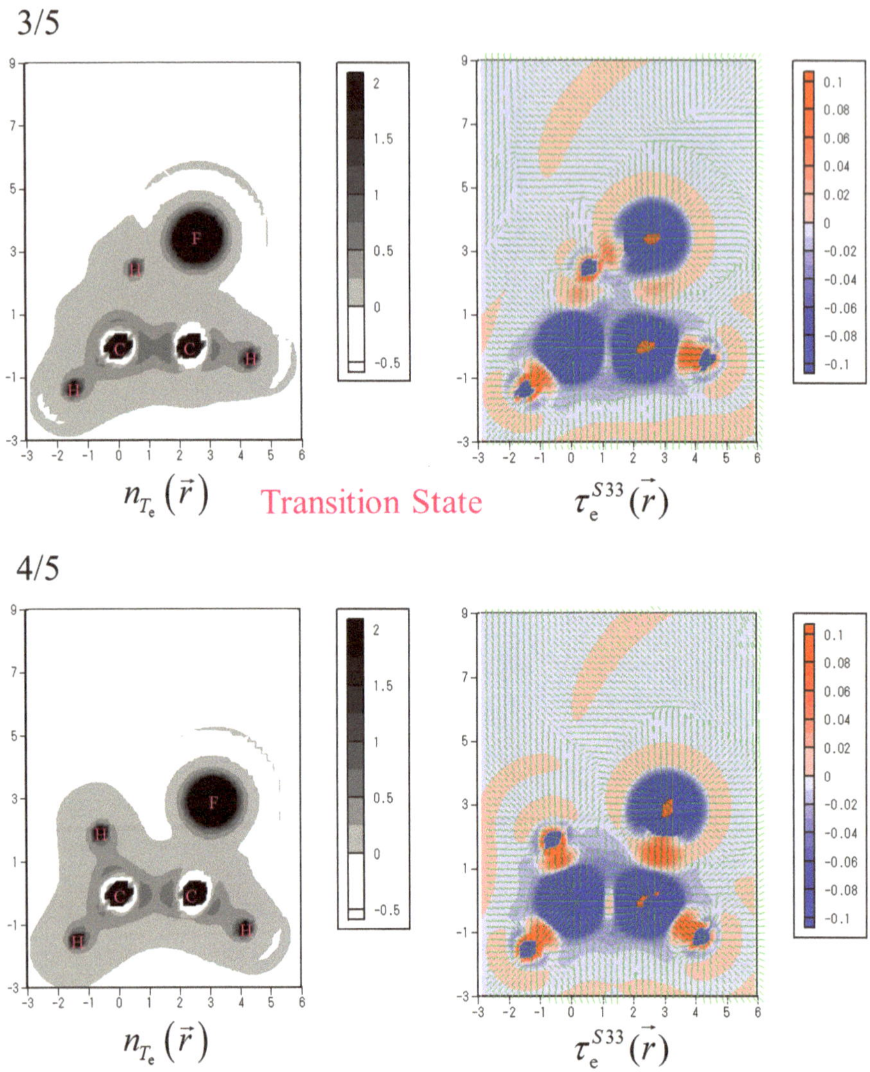

Fig. 3.29 (continued)

occurs in contrast to the degeneracy of two minor eigenvalues of stress in other cases. There appears clear tendency that among all noble gas interactions the largest eigenvalue of stress in Lagrange point becomes negative (see Fig. 3.3). This is because the antibonding orbital contribution overwhelms the bonding one

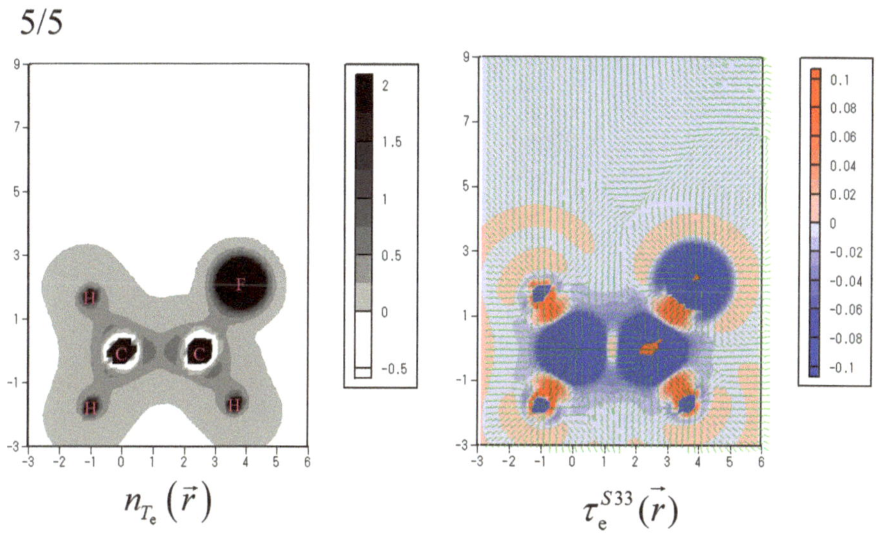

Fig. 3.29 (continued)

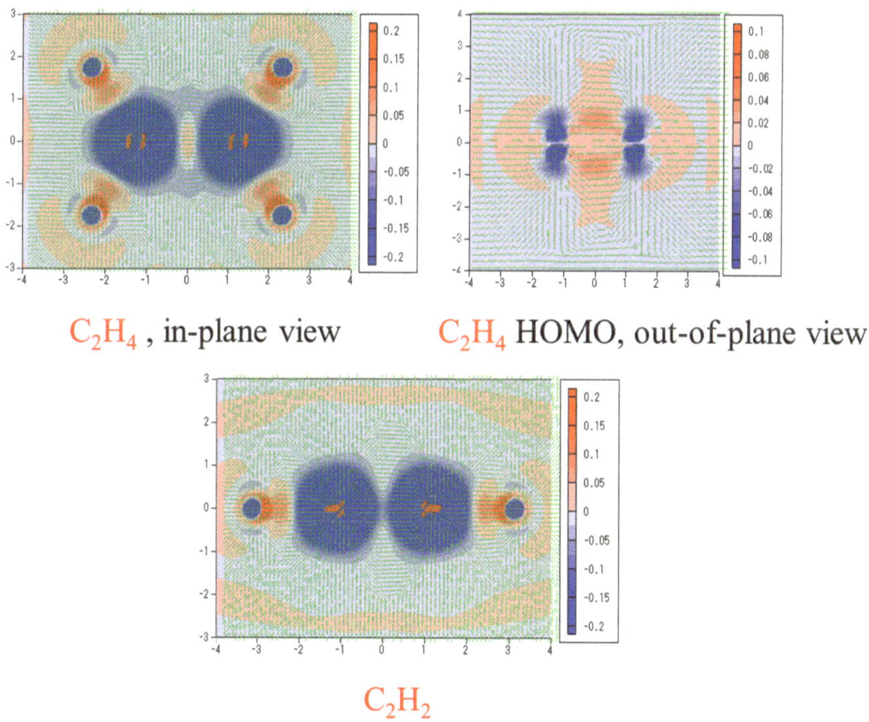

Fig. 3.30 The spindle structures in C_2H_4 and C_2H_2: σ spindle structures of C–C bond and C–H bond together with the sheath structure of π orbital of HOMO (Note that the spindle structure of C–C is immersed in C_2H_2)

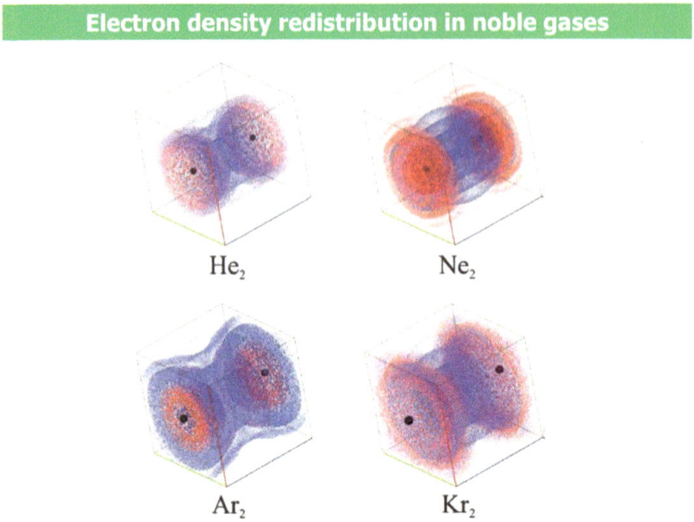

Fig. 3.31 Electron density redistribution in noble gases calculated as difference of molecule and atoms electron densities. The *red dots* show increased electron density; the *blue* dots apply to decreased electron density. The size of dots corresponds to the magnitude of electron density change. The diameter of cube is 20 [a.u.]

(see Fig. 3.20) as realized by the simple orbital interaction scheme (see Fig. 3.4). Actually the electron density is swept away from the Lagrange point (see Fig. 3.31).

There also appears another clear tendency that among all metals and metalloids bonds, the largest eigenvalue of stress in Lagrange point becomes negative. This indicates a kind of fluidity or liquidity (see Figs. 3.9 and 3.10) of bonding/valance electron density between species. There might be a connection between such a feature and band properties of metals and semiconductors however not studied yet. At least, we may say as mentioned in Sect. 3.3.5, the long-range Lewis electron pair formation should emerge. Namely, as the internuclear distance is elongated the spindle structure should be revealed (Ichikawa et al. 2012). Numerical example is Li dimer (see Fig. 3.32).

3.5.2 *Nonclassical Bond Order and Regional Chemical Potential*

The nonclassical bond order concept of energy density and regional chemical potential based bond orders gives natural evaluation of interaction strength compared with classical definition, considering delocalized nature of electrons (Szarek

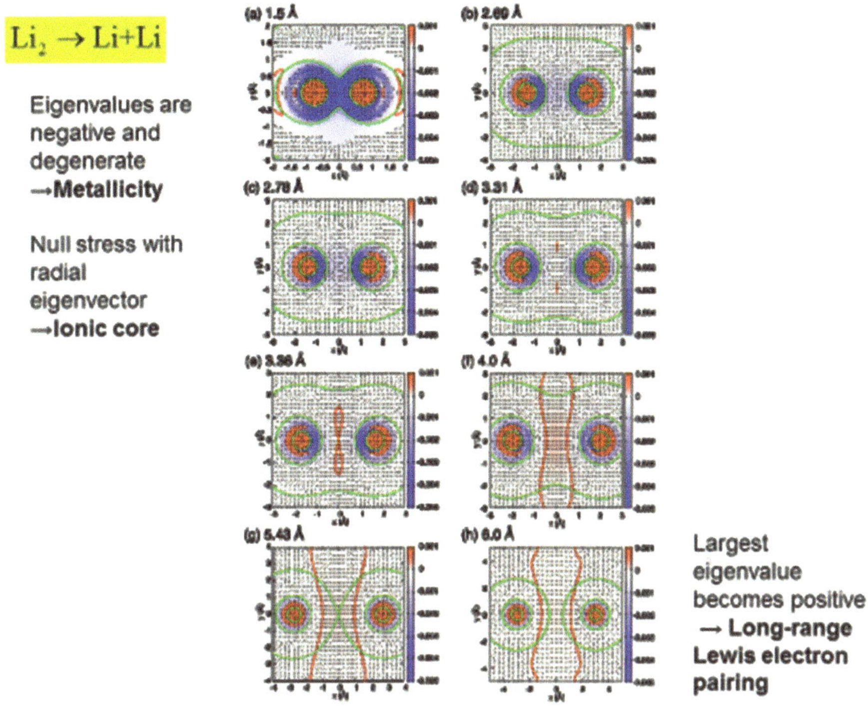

Fig. 3.32 The largest eigenvalue of the stress tensor and corresponding eigenvector of Li_2 at various internuclear distances (**a**) 1.5 Å, (**b**) 2.69 Å (equilibrium distance), (**c**) 2.78 Å, (**d**) 3.31 Å, (**e**) 3.36 Å, (**f**) 4.0 Å, (**g**) 5.43 Å (intrinsic electronic transition state), and (**h**) 6.0 Å

et al. 2008). Numerical examples are organic compounds C_2H_nA with different functional groups (see Fig. 3.33).

Integrated bond order works well for more advanced study of bond strength (Ichikawa et al. 2011). Numerical examples are inverted sandwich-type and open lantern-type dinuclear transition metal complexes (see Fig. 3.34).

The nonclassical bond order (Ichikawa et al. 2014) and the integrated one (Nozaki et al. 2015) have the correlation with the bond force constant. Numerical examples are GeSbTe (GST) alloy, the most popular material for phase change memory (PCM) (see Fig. 3.35).

The local reactivity of hydrogenated Pt clusters has been studied (Szarek et al. 2009). The reaction sites are characterized by lowered electronic regional chemical potential and strong directionality and exhibit electrophilic nature (see Fig. 3.36).

Electronic regional chemical potential work well for the adsorption of Li atoms on the surface of the (12,0) single-wall carbon nanotube (SWCNT) model has been studied (Senami et al. 2011). The adsorption of one lithium atom on the inside of this SWCNT is favored compared to the outside (see Fig. 3.37).

Non-classical bond order

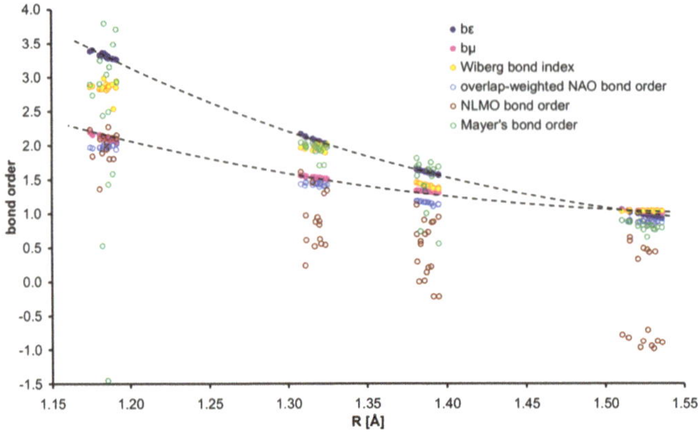

Correlation coefficients

	b_ε	b_μ	b_W	b_{NAO}	b_{NLMO}	b_M
C–C	-0.9884	-0.9931	-0.0099	-0.6456	-0.2049	-0.6352
C=C	-0.9870	-0.9916	-0.0067	-0.4478	0.0673	-0.4172
C≡C	-0.9103	-0.9699	-0.1342	-0.1040	0.1443	-0.0238
C⩦C	-0.9913	-0.9881	-0.9660	-0.8653	-0.1635	-0.0608
In general*	-0.9821	-0.9823	-0.9738	-0.9820	-0.8702	-0.6569

Fig. 3.33 The stress rooted bond orders are *blue* and *pink* dots; NBO bond orders: Wiberg's indices *yellow* dot, atom–atom overlap NAO bond order *blue ring*, NLMO bond orders brown *ring*, and Mayer's bond order *green ring*

Non-classical bond order in open-lantern-type complex

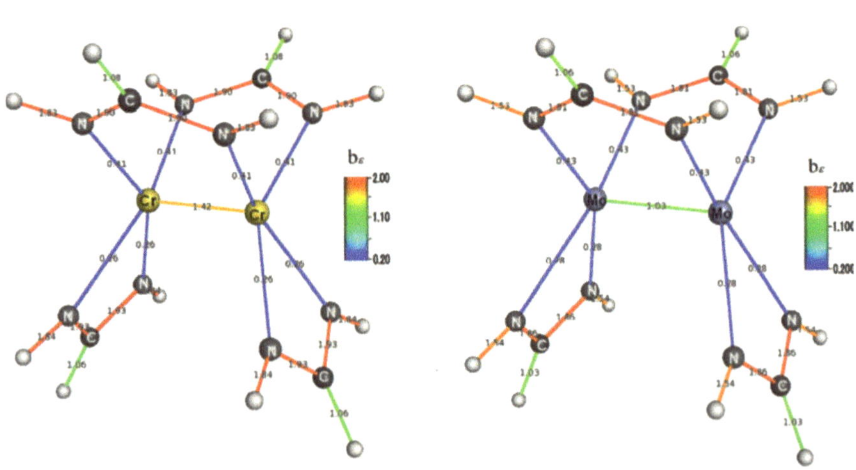

Fig. 3.34 As for the open lantern-type complex, the energy density-based bond order can properly describe the relative strength of Cr–Cr and Mo–Mo bonds by the surface integration of the energy density over the Lagrange surface which can take into account the spatial extent of the orbitals

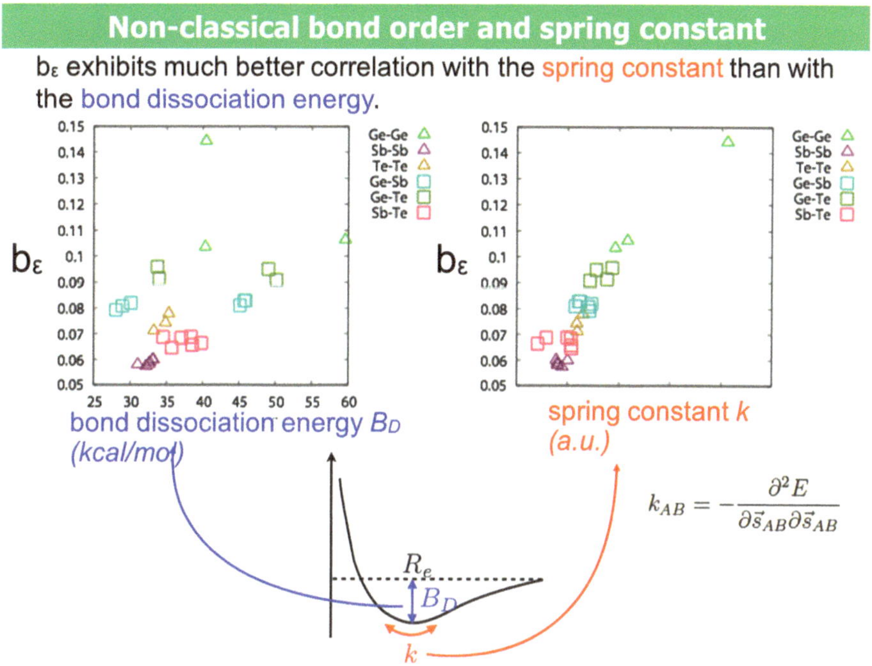

Fig. 3.35 Correlation of the nonclassical bond order and spring constant for GST models for chemical species that may be present in chemical reactions of PCM

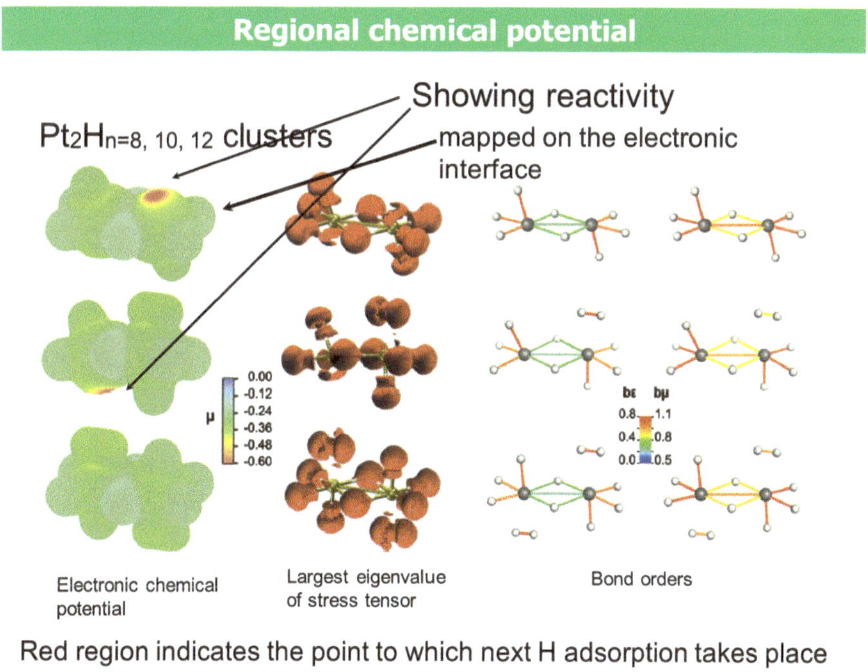

Fig. 3.36 Electronic properties of $Pt_2H_n = 8,10,12$ clusters. From *left* to *right*, electronic chemical potential, largest eigenvalue of stress tensor, and nonclassical bond orders

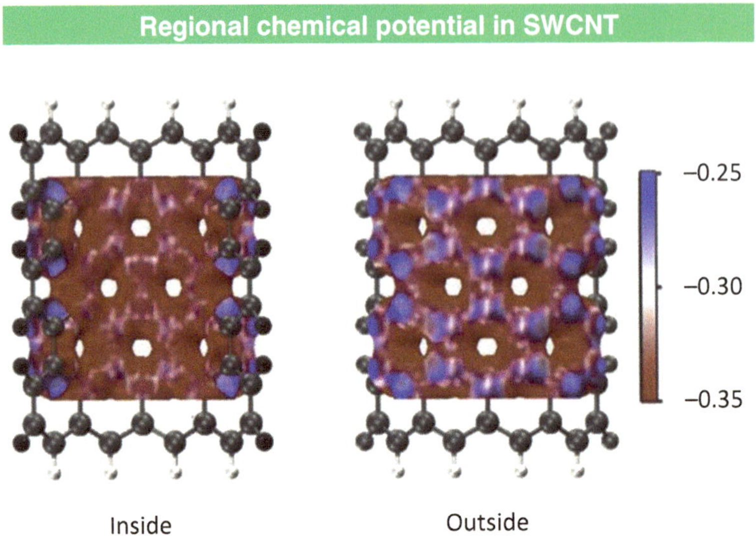

Fig. 3.37 Electronic regional chemical potential in single-wall carbon nanotube (SWCNT) model

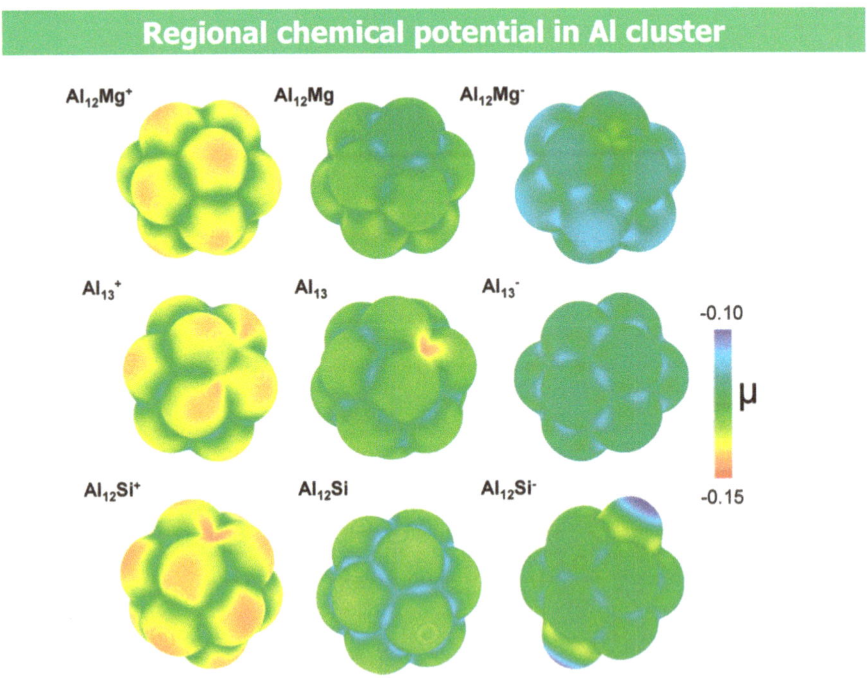

Fig. 3.38 Electronic regional chemical potential for $Al_{12}X^z$ clusters

The nonclassical bond orders and regional chemical potentials work well for the study of the internal bonding, stability, and the regioselectivity of hydrogen on the Al nanoclusters (Henry et al. 2011). Numerical examples are $Al_{12}X^z$ clusters (see Fig. 3.38).

3.6 Summary

The concept of energy density has been developed using stress tensor machineries. The energy density concept has been essential in the quantum field theory, and the stress tensors are used ubiquitously for description of internal forces of matter. Various basic chemical concepts in molecules and chemical reaction systems have been clearly visualized in real three-dimensional space. The new regional energy decomposition scheme has been extended to infinitely small regional energy decomposition scheme, namely, the electronic energy density decomposition scheme.

Not only the short-range force of chemical bonding but also the long-range forces—such as the London force of the nonretarded interaction proportional to R^{-6}, with R being the typical intermolecular distance, and the Casimir–Polder force of the retarded interaction proportional of R^{-7}—are incorporated in this scheme.

Of course, it should be noted that the fact that the adiabatic approximation gives rise to another virial for nuclei, which remains finite for nonequilibrium nuclear configurations. Therefore, we need, as mentioned in Sect. 3.1.1, the wave-packet prescription of the nuclear motion for the simple picture of the regional energy partitioning using Eq. (3.5).

References

J. Bardeen, Phys. Rev. **49**, 653 (1936)
W.A. Bingel, Z. Naturforsch. **18a**, 1249 (1963)
W.A. Bingel, Theoret. Chim. Acta (Berl.) **8**, 54 (1967)
C.A. Coulson, *Valence* (2nd edn., Clarendon, Oxford, 1961)
I. Fleming, *Frontier Orbitals and Organic Chemical Reactions* (Wiley, New York, 1976)
K. Fukui, Acc. Chem. Res. **14**, 363 (1981)
M. Glotzer, Nat. Rev. Mol. Cell Biol. **10**, 9 (2009)
D.J. Henry, P. Szarek, K. Hirai, K. Ichikawa, A. Tachibana, I. Yarovsky, J. Phys. Chem. C **115**, 1714 (2011)
C. Herring, M.H. Nichols, Rev. Mod. Phys. **21**, 185 (1949)
K. Ichikawa, A. Tachibana, Phys. Rev. A **80**, 062507(4) (2009)
K. Ichikawa, A. Wagatsuma, Y. Kurokawa, S. Sakaki, A. Tachibana, Theor. Chem. Accounts **130**, 237 (2011)
K. Ichikawa, H. Nozaki, N. Komazawa, A. Tachibana, AIP Advances **2**, 042195 (16) (2012)
K. Ichikawa, H. Nozaki, A. Tachibana, Mater. Sci. Forum. **783–786**, 2207 (2014)

T. Kato, Commun. Pure Appl. Math. **10**, 151 (1957)
L.D. Landau, E.M. Lifshitz, *Statistical Physics* (Pergamon, Oxford, 1980), pp. 264–265
K. Nakamura, K. Doi, K. Fujitani, A. Tachibana, Phys. Rev. B **71**, 045332(8) (2005)
H. Nozaki, Y. Ikeda, K. Ichikawa, A. Tachibana, J. Comput. Chem. **36**, 1240 (2015)
H. Nozaki, K. Ichikawa, A. Tachibana, Int. J. Quant. Chem. **116**, 504 (2016)
L. Onsager, Phys. Rev. **37**, 405 (1931a)
L. Onsager, Phys. Rev. **38**, 2265 (1931b)
R.T. Pack, W.B. Brown, J. Chem. Phys. **45**, 556 (1966)
L. Pauling, *The Nature of the Chemical Bond*, 3rd edn. (Cornell University, New York, 1960)
L. Pauling, E.B. Wilson Jr., *Introduction to Quantum Mechanics with Application to Chemistry* (McGraw-Hill, New York, 1935)
J.B. Perrin, *Les Atomes* (Liberairie Felix Alcan, 1914)
P. Pyykkö, M. Atsumi, Chem. Eur. J. **15**, 186 (2009)
L.E. Reichl, *A Modern Course in Statistical Physics* (University of Texas, Austin, 1980), pp. 45–47
M. Senami, Y. Ikeda, A. Fukushima, A. Tachibana, AIP Adv. **1**, 042106(12) (2011)
P. Szarek, A. Tachibana, J. Mol. Model. **13**, 651 (2007)
P. Szarek, Y. Sueda, A. Tachibana, J. Chem. Phys. **129**, 094102(16) (2008)
P. Szarek, K. Urakami, C. Zhouz, H. Cheng, A. Tachibana, J. Chem. Phys. **130**, 084111(9) (2009)
A. Tachibana, Phys. Rev. A **35**, 18 (1987a)
A. Tachibana, Int. J. Quant. Chem. S **21**, 181 (1987b)
A. Tachibana, J. Math. Chem. **7**, 95 (1991)
A. Tachibana, in *Conceptual Trends in Quantum Chemistry*, ed by E. S. Kryachko, J. L. Calais. (Kluwer, Dordrecht, 1994), pp. 101–118
A. Tachibana, Int. J. Quant. Chem. **57**, 423 (1996)
A. Tachibana, Theor. Chem. Accounts **102**, 188 (1999a)
A. Tachibana, in *The Transition State – A Theoretical Approach*, ed by T. Fueno. (Kodansha/ Gordon and Breach, Tokyo/Amsterdam, 1999b), pp. 217–247
A. Tachibana, J. Chem. Phys. **115**, 3497 (2001)
A. Tachibana, in *Reviews in Modern Quantum Chemistry:A Celebration of the Contributions of Robert Parr*, ed by K. D. Sen. (World Scientific, Singapore, 2002) Chap.45, pp. 1327–1366
A. Tachibana, in *Fundamental Perspectives in Quantum Chemistry: A Tribute to the Memory of Per-Olov Löwdin*, vol. 2, ed. by E. Brändas, E. Kryachko (Kluwer, Dordrecht, 2003), pp. 211–239
A. Tachibana, Int. J. Quant. Chem. **100**, 981 (2004)
A. Tachibana, J. Mol. Model. **11**, 301 (2005)
A. Tachibana, in *Concepts and Methods in Modern Theoretical Chemistry: Electronic Structure and Reactivity (Atoms, Molecules, and Clusters)*, ed by S. K. Ghosh, P. K. Chattaraj. (Taylor & Francis/CRC, New York, 2013) , Chap.12, pp. 235–251
A. Tachibana, J. Comput. Chem. Jpn. **13**, 18 (2014)
A. Tachibana, T. Iwai, Phys. Rev. A **33**, 2262 (1986)
A. Tachibana, R.G. Parr, Int. J. Quant. Chem. **41**, 527 (1992)
E. Wigner, J. Bardeen, Phys. Rev. **48**, 84 (1935)

Chapter 4
Alpha-Oscillator Theory

Abstract Canonical quantization of QED in finite systems is performed in terms of new b-photon, f-electron, and f^c-positron, which are called alpha-oscillators. The alpha-oscillator algebra is useful for non-perturbationally space-time resolved simulation solving the dual Cauchy problems of the time-dependent QED Hamiltonian $\widehat{H}_{\mathrm{QED}}(t)$. Time-dependent renormalization of QED is performed by using q-number renormalization constants over and above alpha-resonance and thermalization of alpha-oscillators. Quantum mechanics 100 years of mystery on the measurement problem of the Minkowski space-time coordinates is solved.

Keywords Alpha-oscillator energy • Alpha-oscillator theory • Alpha-resonance • Alpha-weighted state • b-photon • Coarse graining • Double slit • Dual Cauchy problem • f-electron • f^c-positron • Einstein–Podolsky–Rosen Measurement • Normal mode • Particle • Q-number renormalization constant • SUGRA • SUSY • Thermalization

4.1 Canonical Quantization

We have an evidence of the accelerated expansion of the universe (see Fig. 4.1). Our universe is not closed but open dynamically. Namely, it is dependent on time. Therefore, it should be remarked that the Hamiltonian of QED could also be dependent on time.

Actually, in harmony with this remark, the QED Hamiltonian is proved to be dependent on time (Tachibana 2016). This time dependence of the QED Hamiltonian does not contradict with that of our universe. It follows that we need time-dependent renormalization of QED for the space-time resolved simulation of molecular dynamics in finite systems. In this chapter, the time-dependent renormalization of QED is performed in terms of the alpha-oscillator theory. The alpha-oscillator theory may furthermore give a natural candidate for now unknown dark energy, which is said to be abundant in our universe and to account for the cause of the accelerated expansion of our universe.

A. Tachibana, *New Aspects of Quantum Electrodynamics*,
DOI 10.1007/978-981-10-3132-8_4

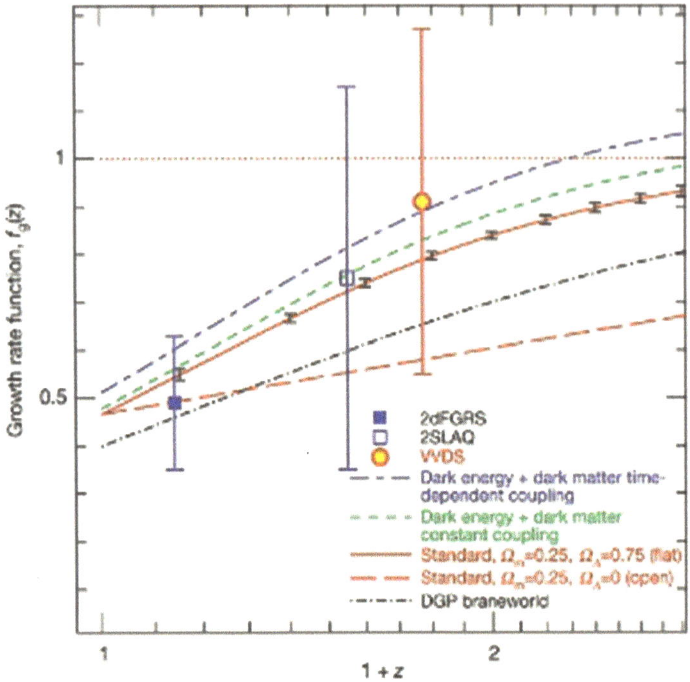

Fig. 4.1 Estimates of the growth rate of cosmic structure compared to predictions from various theoretical models (Reproduced from Guzzo et al. 2008)

4.1.1 QED Hamiltonian

Here in this chapter again, first we start with the Coulomb gauge for the canonical quantization of QED with the conjugate transversal electric field

$$\hat{\vec{E}}_T(x) = -\frac{1}{c}\frac{\partial}{\partial t}\hat{\vec{A}}(x), \quad \mathrm{div}\hat{\vec{A}}(x) = 0 \tag{4.1}$$

The equal-time canonical quantization of the electromagnetic field leads to the equal-time commutation relationships

$$\left[\widehat{A}^i(x), \widehat{A}^j(y)\right]_{x^0=y^0} = 0 \tag{4.2}$$

$$\left[\widehat{E}_T{}^i(x), \widehat{E}_T{}^j(y)\right]_{x^0=y^0} = 0 \tag{4.3}$$

$$\frac{1}{4\pi c}\left[\widehat{A}^i(x), \widehat{E}_T{}^j(y)\right]_{x^0=y^0} = i\hbar\left(\eta^{ij}\delta^3(\vec{x}-\vec{y}) + \partial^i\partial^j\left(-\frac{1}{4\pi}\cdot\frac{1}{|\vec{x}-\vec{y}|}\right)\right) \tag{4.4}$$

Second, the equal-time canonical quantization of the Dirac field leads to the equal-time anti-commutation relationships

$$\{\widehat{\psi}_\ell(x), \widehat{\psi}_{\ell'}(y)\}_{x^0=y^0} = \left\{\widehat{\psi}_\ell{}^\dagger(x), \widehat{\psi}_{\ell'}{}^\dagger(y)\right\}_{x^0=y^0} = 0 \tag{4.5}$$

$$\left\{\widehat{\psi}_\ell(x), \widehat{\psi}_{\ell'}{}^\dagger(y)\right\}_{x^0=y^0} = \delta_{\ell\ell'}\delta^3(\vec{x}-\vec{y}) \tag{4.6}$$

The $\widehat{\psi}(x)$ commutes with $\hat{\vec{A}}(x)$ (Weinberg 1995)

$$\left[\widehat{\psi}(x), \hat{\vec{A}}(x)\right] = 0 \tag{4.7}$$

These fields should of course be renormalized in a step-by-step time-dependent manner, reflecting the time-dependent minimal coupling.

The $\widehat{H}_{\text{QED}}(t)$ is given by using the normal order denoted as : : modulo c-number albeit infinity if any with the obvious notation

$$\widehat{H}_{\text{QED}}(t) = \int d^3\vec{x} : \widehat{H}_{\text{QED}}(x) : \tag{4.8}$$

$$\widehat{H}_{\text{QED}}(x) = \frac{1}{8\pi}\left(\left(\hat{\vec{E}}_T(x)\right)^2 + \left(\text{rot}\hat{\vec{A}}(x)\right)^2\right) - \frac{1}{c}\hat{\vec{j}}(x) \bullet \hat{\vec{A}}(x)$$

$$+\frac{1}{2c}\widehat{j}_0(x)\widehat{A}_0(x) + \widehat{\overline{\psi}}(x)\left(-i\hbar\gamma^k\partial_k + mc\right)\widehat{\psi}(x) \times c \tag{4.9}$$

$$\widehat{A}_0(x) = \int_{-\infty}^{\infty} d^3\vec{y}\,\frac{\widehat{\rho}(y)|_{y^0=x^0}}{|\vec{x}-\vec{y}|} \tag{4.10}$$

At the very beginning, $\widehat{H}_{\text{QED}}(t)$ is defined by integrating the Hamiltonian density operator $\widehat{H}_{\text{QED}}(x)$, a Lorentz scalar under the Poincare symmetry transformation (Weinberg 1995), over the space variables in the Minkowski space-time. Then, it appears that the $\widehat{H}_{\text{QED}}(t)$ is in general dependent on time. There exists the Cauchy problem in order to obtain $\widehat{H}_{\text{QED}}(x)$ for $t > t_0$ (=0) onward with causality and initial condition. With another Cauchy problem for the time evolution of the ket vector with wave function added to, we have the dual Cauchy problem at hand (Tachibana 2015, 2016).

This time-dependent QED Hamiltonian with the dual Cauchy problem fits well with the time-dependent universe.

4.1.2 Conventional Conservative QED Hamiltonian

It should be noted that conventionally we make the QED Hamiltonian independent of time. This is the conventional putative conservative QED Hamiltonian $\widehat{H}_{\text{EMstatic QED}}$

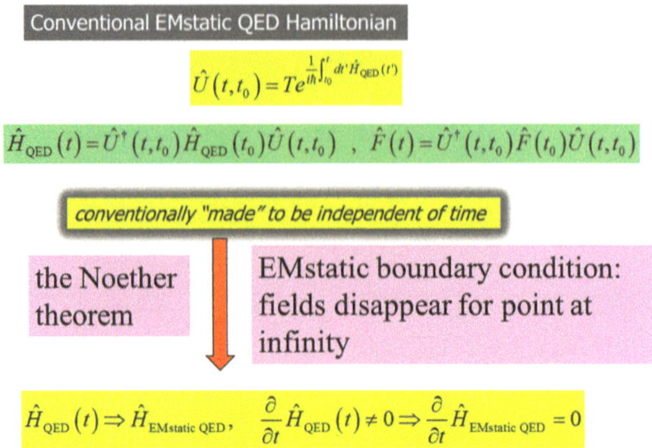

Fig. 4.2 Conventional conservative QED Hamiltonian with the EMstatic field

$$\frac{\partial}{\partial t}\widehat{H}_{\text{QED}} = 0 \text{ if } \widehat{H}_{\text{QED}} = \widehat{H}_{\text{EMstatic QED}} \tag{4.11}$$

with the putative conservative electromagnetostatic (EMstatic) field (see Fig. 4.2).

The basic idea behind this convention is to use the Noether theorem and the putative boundary condition that fields disappear for point at infinity. As far as we follow this convention, the consequence is that the QED Hamiltonian is made to be independent of time, realization of the invariant fields (Landau and Lifshitz 1973) or in other words the putative conservative EMstatic field (see Fig. 4.2).

If at once the putative boundary condition is met, say at the initial time $t = t_0(=0)$, then the putative EMstatic QED Hamiltonian may be used for future time t with $t > t_0(=0)$ onward. Then the time evolution of any field operator $\widehat{F}(x)$ obeys the Heisenberg equation of motion in the well-known form

$$i\hbar\frac{\partial}{\partial t}\widehat{F}(x)\bigg|_{\text{EMstatic QED}} = \left[\widehat{F}(x), \widehat{H}_{\text{EMstatic QED}}\right]\bigg|_{\text{EMstatic QED}} \tag{4.12}$$

with the putative EMstatic QED Hamiltonian $\widehat{H}_{\text{EMstatic QED}}$ with $t > t_0(=0)$ onward. In due course, we need to set up wave function at the initial time $t = t_0(=0)$ in order to discriminate numbers of electrons, positrons, and photons and calculate the expectation value

$$\left\langle \widehat{F}(x) \right\rangle\bigg|_{\text{EMstatic QED}} = \frac{{}_H\langle\Psi|^H\widehat{F}(x)|\Psi\rangle_H}{{}_H\langle\Psi|\Psi\rangle^H H}\Bigg|_{\text{EMstatic QED}} \tag{4.13}$$

where $|\Psi\rangle_H$ denotes the time-independent ket vector in the Heisenberg representation with $t > t_0(=0)$ onward.

Thus, the Cauchy problem of the conventional putative conservative QED Hamiltonian $\widehat{H}_{\text{EMstatic QED}}$ is equivalent to that of quantum mechanics under the putative conservative EMstatic field. It is very simple. Indeed, with initial wave packet given under the putative conservative EMstatic field, the time evolution depends only on the time duration. Namely, if at later time, say $t_i(>t_0)$, and if with the same wave packet given to initiate another event, then exactly the same time evolution should occur. To conclude, if the double-slit phenomenon of Fig. 1.3 be observed in this situation, then it is mystery to realize the stochastic distribution of spots on the screen as Feynman claimed so (Feynman et al. 1965). The reason for the stochastic distribution of spots on the screen is out of order.

This is not the case with the realistic QED Hamiltonian $\widehat{H}_{\mathrm{QED}}(t)$ that is dependent on time. The Noether theorem is still applicable, but we abandon the putative boundary condition that fields disappear for point at infinity. Equivalently, this means that we abandon the putative conservative EMstatic field. With this generic case in mind, the Cauchy problem of fields in QED in the Heisenberg representation has been elaborated elegantly by Nakanishi using ghost field in the Landau gauge (Nakanishi 2004). Here we use the Coulomb gauge, and we do not invoke the additional ghost field. And, we shall apply the alpha-oscillator theory to QED (Tachibana 2016) as elaborated below.

4.2 Alpha-Oscillator Theory

4.2.1 *Synchronization*

To solve for the Cauchy problem of fields in QED, clocks at different space points are synchronized at $t=t_0$, when canonical quantization is performed with the definition of the vacuum ket vector $|0\rangle$. The $\widehat{j}^{\mu}(x)$ develops forward $t>t_0$ with the retarded interactions mediated by photon. The vacuum and field operators are not defined backward $t<t_0$ (see Fig. 4.3).

4.2.2 *Causality and Initial Condition*

To obtain $\widehat{F}(x)$ with $x^{\mu}=(ct,\vec{x})$ in the Minkowski space-time, we may collect information of $\widehat{j}^{\mu}(y)$ with $y^{\mu}=(cu,\vec{y})$ at distant $\vec{y}$ with the retarded time $u=t-\frac{|\vec{x}-\vec{y}|}{c}$ satisfying causality

$$\widehat{j}^{\mu}(cu,\vec{y})=0,\quad u>t \tag{4.14}$$

and initial condition (see Fig. 4.4)

$$\widehat{j}^{\mu}(cu,\vec{y})=0,\quad u<t_0 \tag{4.15}$$

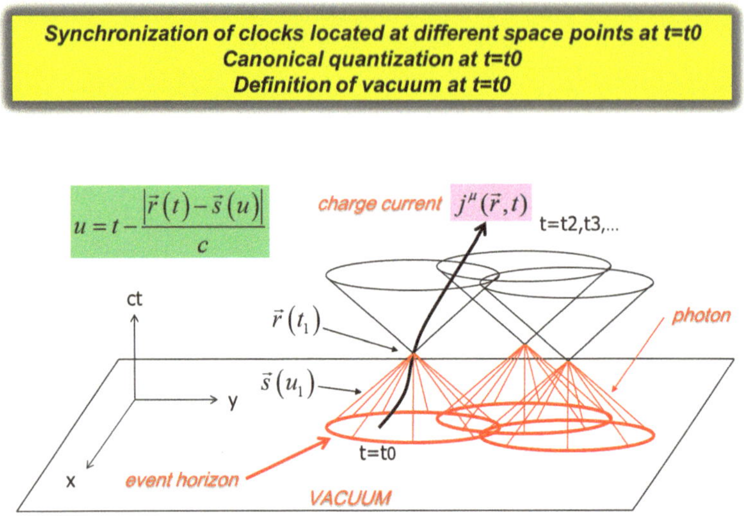

Fig. 4.3 Synchronization of clocks. The charge current develops forward $t > t_0$ with the retarded interactions mediated by photon. The vacuum and field operators are not defined backward $t < t_0$

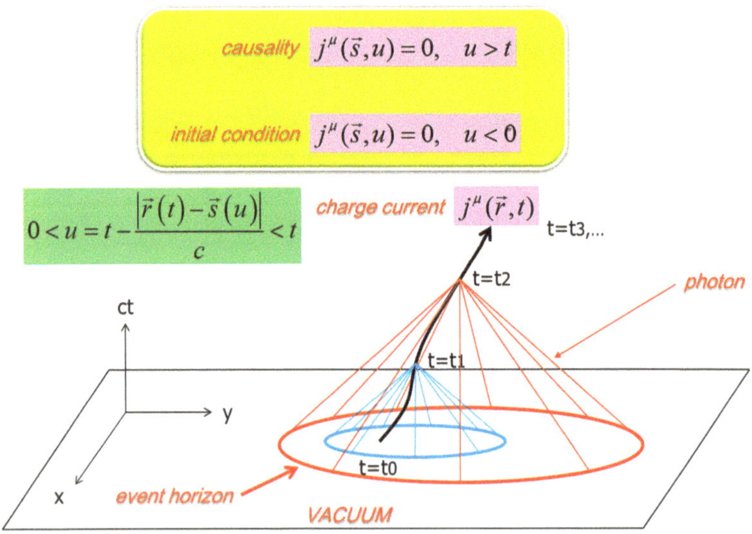

Fig. 4.4 Causality and initial condition

For this purpose, in the following discussions, we may use that any function $F(u)$ satisfying

$$F(u) = 0, \quad u < t_0, \quad u > t \tag{4.16}$$

may be obtained at u with $t_0 < u = t - \frac{|\vec{x}-\vec{y}|}{c} < t$ as (Tachibana 2013)

$$F(u)|_{u=t-\frac{|\vec{x}-\vec{y}|}{c}} = \int_{-\infty}^{\infty} du' F(u')\delta\left(u' - \left(t - \frac{|\vec{x}-\vec{y}|}{c}\right)\right)$$
$$= \frac{|\vec{x}-\vec{y}|}{c\pi}\int_{t_0}^{t} du' \int_{-\infty}^{\infty} d\alpha F(u') e^{i\alpha\left((u'-t)^2 - \frac{(\vec{x}-\vec{y})^2}{c^2}\right)} \tag{4.17}$$

where we have used the delta function

$$\delta\left((u'-t)^2 - a^2\right) = \frac{1}{2a}(\delta((u'-t)-a) + \delta((u'-t)+a)), \quad a > 0 \tag{4.18}$$

with

$$\delta\left((u'-t)^2 - \frac{(\vec{x}-\vec{y})^2}{c^2}\right) = \frac{1}{2\pi}\int_{-\infty}^{\infty} d\alpha e^{i\alpha\left((u'-t)^2 - \frac{(\vec{x}-\vec{y})^2}{c^2}\right)} \tag{4.19}$$

4.2.3 Electromagnetic Field

The vector potential $\hat{\vec{A}}(x)$ should satisfy the Maxwell equation

$$\Box \hat{\vec{A}}(x) = \frac{4\pi}{c}\hat{\vec{j}}_T(x) \tag{4.20}$$

with the transversal charge current

$$\hat{\vec{j}}_T(x) = \hat{\vec{j}}(x) - \frac{1}{4\pi}\,\mathrm{grad}\frac{\partial}{\partial t}\hat{A}_0(x) \tag{4.21}$$

Using the standard Green function, we have (Tachibana 2013)

$$\hat{\vec{A}}(x) = \hat{\vec{A}}_{\mathrm{radiation}}(x) + \hat{\vec{A}}_A(x) \tag{4.22}$$

$$\hat{\vec{A}}_A(ct, \vec{x}) = \frac{1}{c}\int_{-\infty}^{\infty} du \int_{-\infty}^{\infty} d^3\vec{y}\,\frac{\hat{\vec{j}}_{\mathrm{T}}(cu, \vec{y})}{|\vec{x}-\vec{y}|}\delta\left(u - \left(t - \frac{|\vec{x}-\vec{y}|}{c}\right)\right)$$
$$= \frac{1}{c^2\pi}\int_{t_0}^{t} du \int_{-\infty}^{\infty} d\alpha \int_{-\infty}^{\infty} d^3\vec{y}\,\hat{\vec{j}}_{\mathrm{T}}(cu, \vec{y}) e^{i\alpha\left((t-u)^2 - \frac{(\vec{x}-\vec{y})^2}{c^2}\right)} \tag{4.23}$$

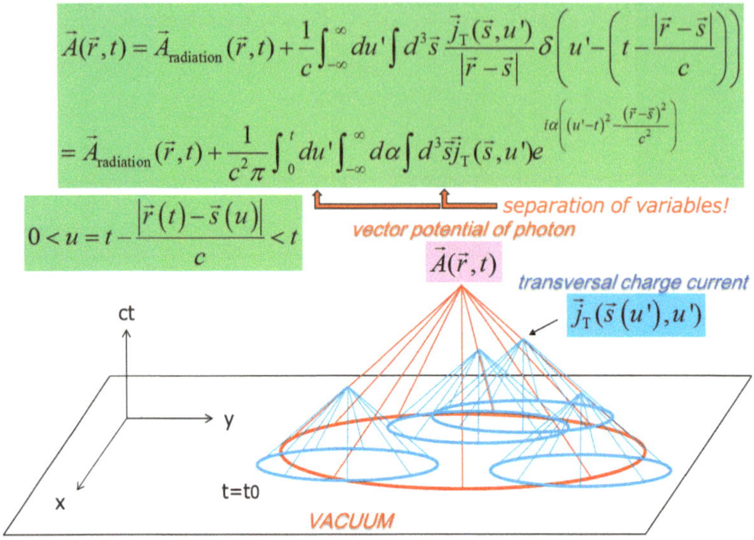

Fig. 4.5 Separation of variables for real-time simulation. Non-causal data are swept out through integration with α

where we omit the contribution of $\hat{\vec{A}}_{\mathrm{M}}(x)$ in Eq. (1.101). It should be noted that we have used the causality and initial condition and then obtained the retarded potential $\hat{\vec{A}}_{A}(x)$ with separation of space-time variables (see Fig. 4.5).

The $\hat{\vec{A}}_{\text{radiation}}(x)$ is given by the $a_{\text{radiation}}$-photon field

$$\hat{\vec{A}}_{\text{radiation}}(x) = \hat{\vec{a}}_{\text{radiation}}(x) + \hat{\vec{a}}^{\dagger}_{\text{radiation}}(x) \tag{4.24}$$

$$\hat{\vec{a}}_{\text{radiation}}(x) = \frac{\sqrt{4\pi\hbar^2 c}}{\sqrt{(2\pi\hbar)^3}} \sum_{\sigma=\pm 1} \int_{-\infty}^{\infty} \frac{d^3\vec{p}}{\sqrt{2p_{\text{radiation}}{}^0}} \hat{a}_{\text{radiation}}(\vec{p},\sigma) e^{-ix_\mu p_{\text{radiation}}{}^{\mu}/\hbar} \vec{e}(\vec{p},\sigma) \tag{4.25}$$

with the usual dispersion relationship of spectrum

$$p_{\text{radiation}}{}^{\mu} = \left(p_{\text{radiation}}{}^{0}, \vec{p}\right), \quad p_{\text{radiation}}{}^{0} = \frac{h\nu_{\text{radiation}}}{c} = |\vec{p}| \tag{4.26}$$

and the polarization vector $\vec{e}(\vec{p},\sigma)$ from Eqs. (1.104), (1.105), (1.106), and (1.107) as

$$\vec{p} \bullet \vec{e}(\vec{p},\sigma) = 0 \tag{4.27}$$

$$\sum_{\sigma=\pm 1} e^i(\vec{p},\sigma)e^{j*}(\vec{p},\sigma) = -\eta^{ij} + \frac{p^i p^j}{-|\vec{p}|^2} \tag{4.28}$$

$$\sum_{i=1}^{3} e^i(\vec{p},\sigma)e^{i*}(\vec{p},\sigma') = \delta_{\sigma\sigma'} \tag{4.29}$$

Note the usual commutation algebra of the $a_{\text{radiation}}$-photon field

$$\left[\hat{a}_{\text{radiation}}(\vec{p},\sigma), \hat{a}_{\text{radiation}}(\vec{q},\sigma')\right] = \left[\hat{a}^{\dagger}{}_{\text{radiation}}(\vec{p},\sigma), \hat{a}^{\dagger}{}_{\text{radiation}}(\vec{q},\sigma')\right] = 0 \tag{4.30}$$

$$\left[\hat{a}_{\text{radiation}}(\vec{p},\sigma), \hat{a}^{\dagger}{}_{\text{radiation}}(\vec{q},\sigma')\right] = \delta_{\sigma\sigma'}\delta^3(\vec{p}-\vec{q}) \tag{4.31}$$

4.2.4 Alpha-Oscillator Algebra

4.2.4.1 Electromagnetic Field

Now with the alpha-oscillator theory applied, the generic solution of the electromagnetic field may be given by using the b-photon field defined as follows (Tachibana 2015)

$$\hat{\vec{A}}(x) = \hat{\vec{b}}(x) + \hat{\vec{b}}^{\dagger}(x) \tag{4.32}$$

$$\hat{\vec{b}}(x) = \frac{\sqrt{4\pi\hbar^2 c}}{\sqrt{(2\pi\hbar)^3}} \sum_{\sigma=\pm 1} \int_0^{\infty} d\nu \int_{-\infty}^{\infty} \frac{d^3\vec{p}}{\sqrt{2p^0(\nu,|\vec{p}|)}} \hat{b}(\nu,\vec{p},\sigma)e^{-i2\pi\nu t}\vec{e}(\vec{p},\sigma)e^{i\vec{x}\cdot\vec{p}/\hbar} \tag{4.33}$$

By using the integral form of the current

$$\hat{\vec{j}}_T(x) = \frac{1}{\sqrt{(2\pi\hbar)^3}} \int_0^{\infty} d\nu \int_{-\infty}^{\infty} d^3\vec{p} \left(\hat{\vec{j}}_T(\nu,\vec{p})e^{-i2\pi\nu t}e^{i\vec{x}\cdot\vec{p}/\hbar} + \hat{\vec{j}}_T^{\dagger}(\nu,\vec{p})e^{+i2\pi\nu t}e^{-i\vec{x}\cdot\vec{p}/\hbar} \right) \tag{4.34}$$

the b-photon field may be represented as

$$\frac{\sqrt{4\pi\hbar^2 c}}{\sqrt{2p^0(\nu,|\vec{p}|)}} \left(-\left(\frac{2\pi\nu}{c}\right)^2 + \frac{|\vec{p}|^2}{\hbar^2} \right) \sum_{\sigma=\pm 1} \hat{b}(\nu,\vec{p},\sigma)\vec{e}(\vec{p},\sigma) = \frac{4\pi}{c} \hat{\vec{j}}_T(\nu,\vec{p}) \tag{4.35}$$

Comparing Eq. (4.35) with Eqs. (4.22), (4.24), and (4.32), we may observe that the $a_{\text{radiation}}$-photon fields are sticking to the b-photon field through $\hat{\vec{j}}_T(x)$. This sticking process may be called “thermalization” of the $a_{\text{radiation}}$-photon fields to the b-photon field. Note that the real positive number $p^0(\nu, |\vec{p}|)$ in Eq. (4.33) is the counterpart of ${p_{\text{radiation}}}^0$ in Eqs. (4.25) and (4.26). The $p^0(\nu, |\vec{p}|)$ is a function of ν and $|\vec{p}|$ serving as the thermalized solution of Eq. (4.35).

The field algebra in Eqs. (4.2), (4.3), and (4.4) are recovered if we assume the b-photon algebra

$$\left[\hat{b}(\nu, \vec{p}, \sigma), \hat{b}(\nu', \vec{q}, \sigma')\right] = \left[\hat{b}^\dagger(\nu, \vec{p}, \sigma), \hat{b}^\dagger(\nu', \vec{q}, \sigma')\right] = 0 \tag{4.36}$$

$$\left[\hat{b}(\nu, \vec{p}, \sigma), \hat{b}^\dagger(\nu', \vec{q}, \sigma')\right] = \delta_{\sigma\sigma'}\delta^3(\vec{p} - \vec{q})\delta(\nu - \nu(|\vec{p}|)_b)\delta(\nu' - \nu(|\vec{q}|)_b) \tag{4.37}$$

where $\nu(|\vec{p}|)_b$ denotes real positive frequency that depends on $|\vec{p}|$. The b-photon field apparently includes the $a_{\text{radiation}}$-photon field in a delta-function form

$$\hat{b}(\nu, \vec{p}, \sigma) \supset \hat{a}_{\text{radiation}}(\vec{p}, \sigma)\delta(\nu - \nu_{\text{radiation}}) \tag{4.38}$$

Then, the electromagnetic part of $\hat{H}_{QED}(t)$ (modulo c-number vacuum energy) in Eqs. (4.8) and (4.9) is given as

$$\begin{aligned}
\hat{H}_{\text{QED}}(t) &\supset \int d^3\vec{x} : \frac{1}{8\pi}\left(\left(\hat{\vec{E}}_T(x)\right)^2 + \left(\text{rot}\hat{\vec{A}}(x)\right)^2\right) : \\
&= \hbar^2 c \sum_{\sigma=\pm 1} \int_0^\infty d\nu \int_0^\infty d\nu' \int_{-\infty}^\infty \frac{d^3\vec{p}}{\sqrt{2p^0(\nu, |\vec{p}|)}\sqrt{2p^0(\nu', |\vec{p}|)}} \\
&\times \left(\left(\frac{2\pi\nu}{c}\right)\left(\frac{2\pi\nu'}{c}\right) + \frac{|\vec{p}|^2}{\hbar^2}\right)\hat{b}^\dagger(\nu, \vec{p}, \sigma)\hat{b}(\nu', \vec{p}, \sigma)e^{i2\pi(\nu-\nu')t} \\
&\text{(modulo c-number)}
\end{aligned} \tag{4.39}$$

which part may depend on t and t_0 since $\hat{H}_{QED}(t)$ is dependent on t and t_0. Moreover, Eq. (4.39) includes the radiation part (modulo time-independent c-number vacuum energy) given as

$$\begin{aligned}
&\int d^3\vec{x} : \frac{1}{8\pi}\left(\left(\hat{\vec{E}}_T(x)\right)^2 + \left(\text{rot}\hat{\vec{A}}(x)\right)^2\right) : \\
&\supset \int d^3\vec{x} : \frac{1}{8\pi}\left(\left(\hat{\vec{E}}_{T_{\text{radiation}}}(x)\right)^2 + \left(\text{rot}\hat{\vec{A}}_{\text{radiation}}(x)\right)^2\right) : \\
&= \sum_{\sigma=\pm 1} \int_{-\infty}^\infty d^3\vec{p}c{p_{\text{radiation}}}^0{\hat{a}_{\text{radiation}}}^\dagger(\vec{p}, \sigma)\hat{a}_{\text{radiation}}(\vec{p}, \sigma) \\
&\text{(modulo time-independent c-number)}
\end{aligned} \tag{4.40}$$

which is manifestly independent of t as well as t_0.

4.2.4.2 The Dirac Field

The $\widehat{\psi}(x)$ may be given by using another standard Green function $K(x,y)$ as (Weinberg 1995)

$$\widehat{\psi}(x) = \widehat{\psi}_{\text{free}}(x) + \frac{1}{i\hbar}\int d^4y K(x,y)\left(-\frac{q}{c}\widehat{A\!\!\!/}(y)\right)\widehat{\psi}(y) \tag{4.41}$$

$$\left(-i\hbar\partial\!\!\!/ + mc\right)K(x,y) = i\hbar\delta^4(x-y) \tag{4.42}$$

where $\widehat{\psi}_{\text{free}}(x)$ denotes the free field. The $\widehat{\psi}_{\text{free}}(x)$ is given by the free e_{free}-electron and $e_{\text{free}}{}^{c}$-positron fields

$$\widehat{\psi}_{\text{free}}(x) = \widehat{e}_{\text{free}}(x) + \widehat{e}_{\text{free}}{}^{c\dagger}(x) \tag{4.43}$$

$$\widehat{e}_{\text{free}_\ell}(x) = \frac{1}{\sqrt{(2\pi\hbar)^3}}\sum_{\sigma=\pm\frac{1}{2}}\int_{-\infty}^{\infty} d^3\vec{p}\,\widehat{e}_{\text{free}}(\vec{p},\sigma)e^{-ix_\mu p_{\text{free}}{}^{\mu}/\hbar}u_\ell(\vec{p},\sigma) \tag{4.44}$$

$$\widehat{e}_{\text{free}_\ell}{}^{c\dagger}(x) = \frac{1}{\sqrt{(2\pi\hbar)^3}}\sum_{\sigma=\pm\frac{1}{2}}\int_{-\infty}^{\infty} d^3\vec{p}\,\widehat{e}_{\text{free}}{}^{c\dagger}(\vec{p},\sigma)e^{+ix_\mu p_{\text{free}}{}^{\mu}/\hbar}v_\ell(\vec{p},\sigma) \tag{4.45}$$

with the usual dispersion relationship of spectrum

$$p_{\text{free}}{}^{\mu} = \left(p_{\text{free}}{}^{0},\vec{p}\right),\quad p_{\text{free}}{}^{0} = \frac{h\nu_{\text{free}}}{c} = \sqrt{(mc)^2 + |\vec{p}|^2} \tag{4.46}$$

and the anti-commutation algebra

$$\begin{aligned}\{\widehat{e}_{\text{free}}(\vec{p},\sigma),\widehat{e}_{\text{free}}(\vec{q},\sigma')\} &= \{\widehat{e}_{\text{free}}{}^{c}(\vec{p},\sigma),\widehat{e}_{\text{free}}{}^{c}(\vec{q},\sigma')\}\\ &= \left\{\widehat{e}_{\text{free}}{}^{\dagger}(\vec{p},\sigma),\widehat{e}_{\text{free}}{}^{\dagger}(\vec{q},\sigma')\right\} = \{\widehat{e}_{\text{free}}{}^{c\dagger}(\vec{p},\sigma),\widehat{e}_{\text{free}}{}^{c\dagger}(\vec{q},\sigma')\} = 0\end{aligned} \tag{4.47}$$

$$\left\{\widehat{e}_{\text{free}}(\vec{p},\sigma),\widehat{e}_{\text{free}}{}^{\dagger}(\vec{q},\sigma')\right\} = \left\{\widehat{e}_{\text{free}}{}^{c}(\vec{p},\sigma),\widehat{e}_{\text{free}}{}^{c\dagger}(\vec{q},\sigma')\right\} = \delta_{\sigma\sigma'}\delta^3(\vec{p}-\vec{q}) \tag{4.48}$$

The Dirac spinors $u(\vec{p},\sigma)$ for electron and $v(\vec{p},\sigma)$ for positron satisfy

$$\left(p_{\text{free}}{}^{\mu}\gamma_\mu - mc\right)u(\vec{p},\sigma) = 0 \tag{4.49}$$

$$\left(p_{\text{free}}{}^{\mu}\gamma_\mu + mc\right)v(\vec{p},\sigma) = 0 \tag{4.50}$$

$$\sum_{\sigma=\pm\frac{1}{2}} u(\vec{p},\sigma)\bar{u}(\vec{p},\sigma) = \frac{1}{2p_{\text{free}}{}^{0}}\left(p_{\text{free}}{}^{\mu}\gamma_\mu + mc\right) \tag{4.51}$$

$$\sum_{\sigma=\pm\frac{1}{2}} v(\vec{p},\sigma)\bar{v}(\vec{p},\sigma) = \frac{1}{2p_{\text{free}}{}^{0}}\left(p_{\text{free}}{}^{\mu}\gamma_\mu - mc\right) \tag{4.52}$$

$$\bar{u}(\vec{p},\sigma)\gamma^{\mu}u(\vec{p},\sigma') = \bar{v}(\vec{p},\sigma)\gamma^{\mu}v(\vec{p},\sigma') = \left(p_{\text{free}}{}^{\mu}/p_{\text{free}}{}^{0}\right)\delta_{\sigma\sigma'} \tag{4.53}$$

$$\bar{u}(\vec{p},\sigma)\gamma^{0}v(-\vec{p},\sigma') = \bar{v}(\vec{p},\sigma)\gamma^{0}u(-\vec{p},\sigma') = 0 \tag{4.54}$$

Now with the alpha-oscillator theory applied, the generic solution may be given by using the$f$-electron and f^{c}-positron fields defined as follows:

$$\widehat{\psi}(x) = \widehat{f}(x) + \widehat{f}^{c\dagger}(x) \tag{4.55}$$

$$\widehat{f}_{\ell}(x) = \frac{1}{\sqrt{(2\pi\hbar)^3}}\sum_{\sigma=\pm\frac{1}{2}}\int_0^{\infty}d\nu\int_{-\infty}^{\infty}d^3\vec{p}\widehat{f}(\nu,\vec{p},\sigma)e^{-i2\pi\nu t}u_{\ell}(\vec{p},\sigma)e^{i\vec{x}\bullet\vec{p}/\hbar} \tag{4.56}$$

$$\widehat{f^{c}}_{\ell}{}^{\dagger}(x) = \frac{1}{\sqrt{(2\pi\hbar)^3}}\sum_{\sigma=\pm\frac{1}{2}}\int_0^{\infty}d\nu\int_{-\infty}^{\infty}d^3\vec{p}\widehat{f}^{c\dagger}(\nu,\vec{p},\sigma)e^{+i2\pi\nu t}v_{\ell}(\vec{p},\sigma)e^{-i\vec{x}\bullet\vec{p}/\hbar} \tag{4.57}$$

Applying the first thermalization of the b-photon field Eq. (4.35) to the Dirac Eq. (1.16), we obtain the second thermalization of the f-electron field

$$\begin{aligned}
&\frac{q}{c}\gamma^{0}\int_0^{\infty}d\nu'\int_{-\infty}^{\infty}d^3\vec{q}\widehat{A}_0(\nu-\nu',\vec{p}-\vec{q})\sum_{\sigma=\pm\frac{1}{2}}\widehat{f}(\nu',\vec{q},\sigma)u(\vec{q},\sigma)\\
&=\frac{4\pi}{c}\int_0^{\infty}d\nu'\int_{-\infty}^{\infty}\frac{d^3\vec{q}}{\left(-\left(\frac{2\pi\nu'}{c}\right)^2+\frac{|\vec{q}|^2}{\hbar^2}\right)}\Bigg(\gamma_k\widehat{j}_T{}^{k}(\nu',\vec{q})\sum_{\sigma=\pm\frac{1}{2}}\widehat{f}(\nu-\nu',\vec{p}-\vec{q},\sigma)u(\vec{p}-\vec{q},\sigma)\\
&\quad+\gamma_k\widehat{j}_T{}^{\dagger k}(\nu',\vec{q})\sum_{\sigma=\pm\frac{1}{2}}\widehat{f}(\nu+\nu',\vec{p}+\vec{q},\sigma)u(\vec{p}+\vec{q},\sigma)\Bigg)
\end{aligned} \tag{4.58}$$

with

$$\begin{aligned}
\widehat{A}_0(\nu,\vec{p}) = {}& \frac{q}{(2\pi\hbar)^3}\sum_{\sigma=\pm\frac{1}{2}}\sum_{\sigma'=\pm\frac{1}{2}}\int_0^{\infty}d\nu'\int_{-\infty}^{\infty}d^3\vec{q}\\
&\times\Big(\widehat{f}^{\dagger}(\nu',\vec{q},\sigma)\widehat{f}(\nu+\nu',\vec{p}+\vec{q},\sigma')u^{\dagger}(\vec{q},\sigma)u(\vec{p}+\vec{q},\sigma')\\
&+\widehat{f}^{\dagger}(\nu',\vec{q},\sigma)\widehat{f}^{c\dagger}(-\nu-\nu',-\vec{p}-\vec{q},\sigma')u^{\dagger}(\vec{q},\sigma)v(-\vec{p}-\vec{q},\sigma')\\
&+\widehat{f}^{c}(\nu',\vec{q},\sigma)\widehat{f}(\nu-\nu',\vec{p}-\vec{q},\sigma')v^{\dagger}(\vec{q},\sigma)u(\vec{p}-\vec{q},\sigma')\\
&+\widehat{f}^{c}(\nu',\vec{q},\sigma)\widehat{f}^{c\dagger}(-\nu+\nu',-\vec{p}+\vec{q},\sigma')v^{\dagger}(\vec{q},\sigma)v(-\vec{p}+\vec{q},\sigma')\Big)
\end{aligned} \tag{4.59}$$

and the third thermalization of the f^c-positron field

$$\frac{q}{c}\gamma^0\int_0^{\infty}d\nu'\int_{-\infty}^{\infty}d^3\vec{q}\widehat{A}_0^{\dagger}(\nu-\nu',\vec{p}-\vec{q})\sum_{\sigma=\pm\frac{1}{2}}\widehat{f}^{c\dagger}(\nu',\vec{q},\sigma)v(\vec{q},\sigma)$$
$$=\frac{4\pi}{c}\int_0^{\infty}d\nu'\int_{-\infty}^{\infty}\frac{d^3\vec{q}}{\left(-\left(\frac{2\pi\nu'}{c}\right)^2+\frac{|\vec{q}|^2}{\hbar^2}\right)}\Bigg(\gamma_k\widehat{j}_T{}^{k}(\nu',\vec{q})\sum_{\sigma=\pm\frac{1}{2}}\widehat{f}^{c\dagger}(\nu+\nu',\vec{p}+\vec{q},\sigma)$$
$$\times v(\vec{p}+\vec{q},\sigma)+\gamma_k\widehat{j}_T{}^{\dagger k}(\nu',\vec{q})\sum_{\sigma=\pm\frac{1}{2}}\widehat{f}^{c\dagger}(\nu-\nu',\vec{p}-\vec{q},\sigma)\times v(\vec{p}-\vec{q},\sigma)\Bigg) \tag{4.60}$$

The field algebra in Eqs. (4.5) and (4.6) are recovered if we assume the *f*-electron and f^c-positron algebras

$$\begin{aligned}&\left\{\widehat{f}(\nu,\vec{p},\sigma),\widehat{f}(\nu',\vec{q},\sigma')\right\}=0,\quad\left\{\widehat{f}^c(\nu,\vec{p},\sigma),\widehat{f}^c(\nu',\vec{q},\sigma')\right\}=0\\&\left\{\widehat{f}^{\dagger}(\nu,\vec{p},\sigma),\widehat{f}^{\dagger}(\nu',\vec{q},\sigma')\right\}=0,\quad\left\{\widehat{f}^{c\dagger}(\nu,\vec{p},\sigma),\widehat{f}^{c\dagger}(\nu',\vec{q},\sigma')\right\}=0\end{aligned} \tag{4.61}$$

$$\begin{aligned}&\left\{\widehat{f}(\nu,\vec{p},\sigma),\widehat{f}^{\dagger}(\nu',\vec{q},\sigma')\right\}=\delta_{\sigma\sigma'}\delta^3(\vec{p}-\vec{q})\delta\Big(\nu-\nu(|\vec{p}|)_f\Big)\delta\Big(\nu'-\nu(|\vec{q}|)_f\Big),\\&\left\{\widehat{f}^c(\nu,\vec{p},\sigma),\widehat{f}^{c\dagger}(\nu',\vec{q},\sigma')\right\}=\delta_{\sigma\sigma'}\delta^3(\vec{p}-\vec{q})\delta\Big(\nu-\nu(|\vec{p}|)_f\Big)\delta\Big(\nu'-\nu(|\vec{q}|)_f\Big)\end{aligned} \tag{4.62}$$

where $\nu(|\vec{p}|)_f$ denotes real positive frequency that depends on $|\vec{p}|$. Also, Eq. (4.7) is recovered if we assume

$$\begin{aligned}\left[\widehat{f}(\nu,\vec{p},\sigma),\widehat{b}(\nu',\vec{q},\sigma')\right]&=\left[\widehat{f}^c(\nu,\vec{p},\sigma),\widehat{b}(\nu',\vec{q},\sigma')\right]\\&=\left[\widehat{f}^{\dagger}(\nu,\vec{p},\sigma),\widehat{b}(\nu',\vec{q},\sigma')\right]\\&=\left[\widehat{f}^{c\dagger}(\nu,\vec{p},\sigma),\widehat{b}(\nu',\vec{q},\sigma')\right]=0\end{aligned} \tag{4.63}$$

The *f*-electron and f^c-positron fields apparently include the e_{free}-electron and $e_{\text{free}}{}^{c}$-positron fields, respectively, in the delta-function forms

$$\widehat{f}(\nu,\vec{p},\sigma)\supset\widehat{e}_{\text{free}}(\vec{p},\sigma)\delta(\nu-\nu_{\text{free}}) \tag{4.64}$$

$$\widehat{f}^c(\nu,\vec{p},\sigma) \supset \widehat{e}_{\text{free}}{}^c(\vec{p},\sigma)\delta(\nu-\nu_{\text{free}}) \tag{4.65}$$

Then, the Dirac part of $\widehat{H}_{\text{QED}}$ (modulo c-number vacuum energy) in Eqs. (4.8) and (4.9) is given as

$$\begin{aligned}\widehat{H}_{\text{QED}}(t) &\supset \int d^3\vec{x} : \widehat{\bar{\psi}}(x)\left(-i\hbar\gamma^k\partial_k + mc\right)\widehat{\psi}(x)\times c : \\ &= \sum_{\sigma=\pm\frac{1}{2}}\int_0^\infty d\nu\int_0^\infty d\nu'\int_{-\infty}^\infty d^3\vec{p}cp_{\text{free}}{}^0 \times \left(\widehat{f}^\dagger(\nu\vec{p}\sigma)\widehat{f}(\nu',\vec{p},\sigma)e^{+i2\pi(\nu-\nu')t}\right.\\ &\left.+\widehat{f}^{c\dagger},(\nu',\vec{p},\sigma),\widehat{f}^c,(\nu,\vec{p},\sigma),e^{-i2\pi(\nu-\nu')t}\right)(\text{modulo c-number})\end{aligned} \tag{4.66}$$

which part may depend on t and t_0 since $\hat{H}_{QED}(t)$ is dependent on t and t_0. Moreover, Eq. (4.66) includes the free part (modulo time-independent c-number vacuum energy) given as

$$\begin{aligned}&\int d^3\vec{x} : \widehat{\bar{\psi}}(x)\left(-i\hbar\gamma^k\partial_k + mc\right)\widehat{\psi}(x)\times c : \\ &\supset \int d^3\vec{x} : \widehat{\bar{\psi}}_{\text{free}}(x)\left(-i\hbar\gamma^k\partial_k + mc\right)\widehat{\psi}_{\text{free}}(x)\times c : \\ &= \sum_{\sigma=\pm\frac{1}{2}}\int_{-\infty}^\infty d^3\vec{p}cp_{\text{free}}{}^0\left(\widehat{e}_{\text{free}}{}^\dagger(\vec{p},\sigma)\widehat{e}_{\text{free}}(\vec{p},\sigma)+\widehat{e}_{\text{free}}{}^{c\dagger}(\vec{p},\sigma)\widehat{e}_{\text{free}}{}^c(\vec{p},\sigma)\right)\\ &(\text{modulo time-independent c-number})\end{aligned} \tag{4.67}$$

which is manifestly independent of t as well as t_0.

4.2.4.3 Alpha-Oscillator with Resonance and Thermalization

We have performed quantization of QED using new sub-particles, b-photon, f-electron, and f^c-positron, which are called alpha-oscillators. The particle fields of photon, electron, and positron are constructed by infinite superposition of those corresponding fields of alpha-oscillators.

The alpha-oscillators are functions of ω denoting the collected set of variables of ν-frequency, $\vec{p}$-momentum, and σ-spin

$$\omega = \{\nu,\vec{p},\sigma\} \tag{4.68}$$

A real positive number $p^0(\nu, |\vec{p}|)$ is a function of ν and $|\vec{p}|$. For any $\vec{p}$-momentum, there exists a particular ν-frequency $\nu(|\vec{p}|)$ and satisfies

$$p^0(\nu(|\vec{p}|), |\vec{p}|) = \frac{h\nu(|\vec{p}|)}{c} \tag{4.69}$$

which is called the alpha-resonance condition. The operator dynamics of the alpha-oscillators has been formulated in terms of thermalization; see Eq. (4.35) for photon and Eqs. (4.60), (4.61), and (4.62) for electron and positron.

4.3 Double-Slit Space-Time-Resolved Prediction of QED

4.3.1 The Feynman Mystery

Let us ask an apparently mysterious question: what causes the time-dependent randomness of the sequential spots observed in the buildup of electron double-slit interference pattern, an experiment performed by Tonomura (2005)? It is mysterious as so mentioned by Feynman et al. (1965). We should admit the mystery if we rely on quantum mechanics and even on QED under the conventional putative time-independent EMstatic field (see Sect. 4.1.2). But if the Hamiltonian were time dependent, then it might cause the time-dependent randomness of the sequential spots. Then the next "consequent" question is do we have such reasonable time-dependent Hamiltonian in reality? Yes, $\widehat{H}_{\mathrm{QED}}(t)$ is the key to answer the question (see Fig. 4.6).

Now that we have $\widehat{H}_{\mathrm{QED}}(t)$ at hand (see Fig. 4.7), for an event α_i starting at t_i with $t_0 < t_i;\ i = 1, 2, 3, \ldots$, we set up the initial ket vector associated with the wave

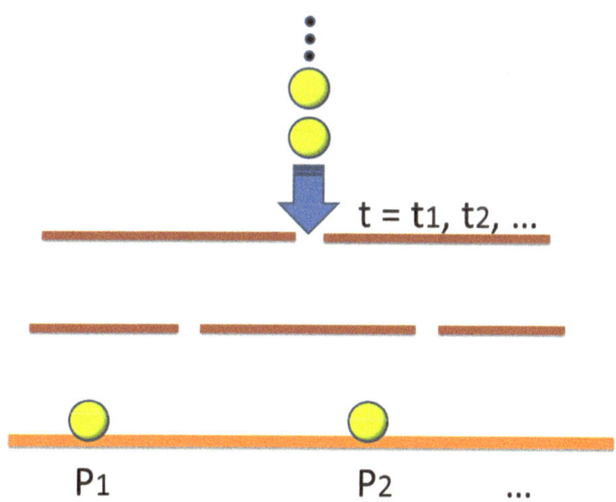

Fig. 4.6 Double-slit space-time resolved prediction of QED using the time-dependent Hamiltonian $\widehat{H}_{\mathrm{QED}}(t)$ rather than the conventional putative time-independent EMstatic Hamiltonian $\widehat{H}_{\mathrm{EMstatic\ QED}}$ or the conventional time-independent electrostatic Hamiltonian $\widehat{H}_{\mathrm{electrostatic}}$ of quantum mechanics

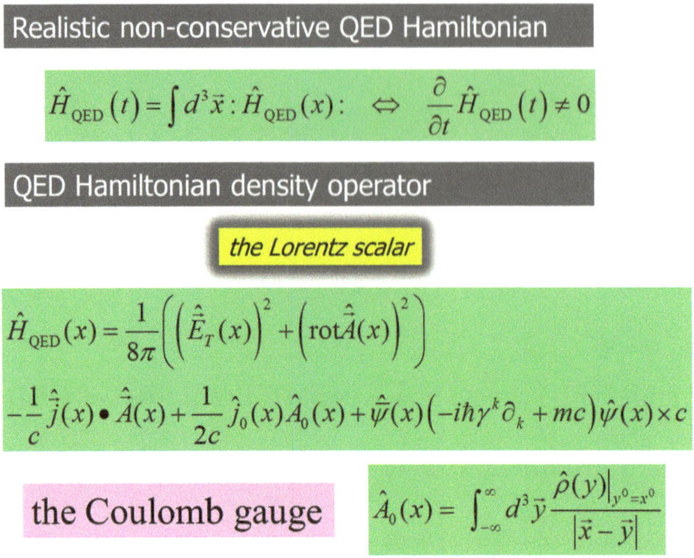

Fig. 4.7 Realistic nonconservative QED Hamiltonian at hand

function. For example, first of all for electron (or positron), the phenomenon at t_{Pi} following t_i is not the collapse of wave packet but the reaction of electron (or positron) with material at the screen. Second, for photon, the phenomenon at t_{Pi} following t_i is not the collapse of wave packet but the formation of electronic excited state, say exciton pair, on the screen, where the photon disappears as shown in Fig. 4.8 (Tachibana 2016). The combination of the particle number nonconservation and the exciton pair formation associated with the time-dependent $\widehat{H}_{\mathrm{QED}}(t)$is the real phenomenon (see Fig. 4.8).

4.3.2 *The Dual Cauchy Problem*

The time evolutions of fields and ket vectors with wave functions in QED constitute the dual Cauchy problem. The unified treatment is given as follows.

4.3.2.1 Time Evolution Operator

The time evolution of q-number $\widehat{F}(t)(= \widehat{F}(ct, \vec{x}))$ for $t > t_0$ $(=0)$ onward is brought about by the time evolution operator $\widehat{U}(t, t_0)$ obeying

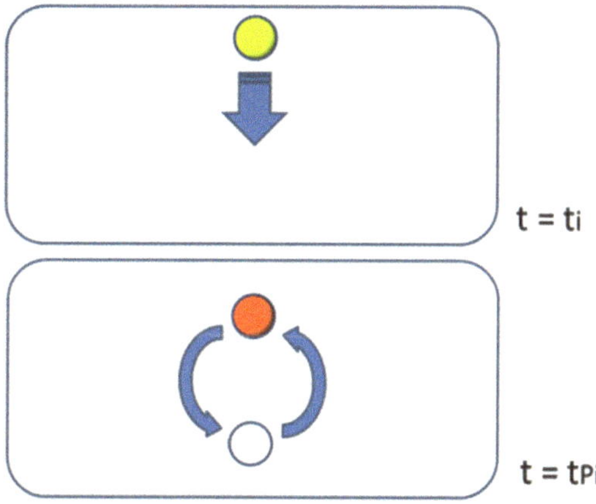

Fig. 4.8 Photon number nonconservation: space-time-resolved prediction in QED with time-dependent $\widehat{H}_{\mathrm{QED}}(t)$. The conventional putative time-independent EMstatic Hamiltonian $\widehat{H}_{\mathrm{EMstatic\ QED}}$ fails to account for the photon number nonconservation. Nor in quantum mechanics also, every particle number is conserved, so that mysterious "collapse of wave packet" should be invoked.

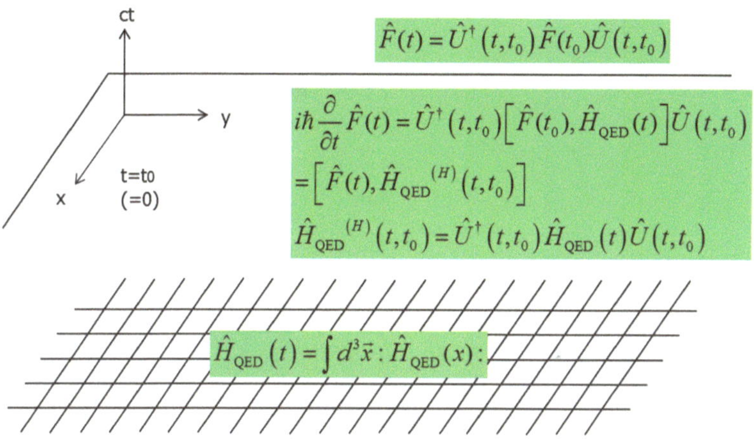

Fig. 4.9 The time evolution of q-number $\widehat{F}(t)\left(= \widehat{F}(ct, \vec{x})\right)$ using the time evolution operator $\widehat{U}(t, t_0)$ with $t > t_0$ onward

$$i\hbar\frac{\partial}{\partial t}\widehat{U}(t,t_0) = \widehat{H}_{\mathrm{QED}}(t)\widehat{U}(t,t_0), \quad i\hbar\frac{\partial}{\partial t}\widehat{U}^{\dagger}(t,t_0) = -\widehat{U}^{\dagger}(t,t_0)\widehat{H}_{\mathrm{QED}}(t) \tag{4.70}$$

and the solution (see Fig. 4.9)

$$\widehat{U}(t,t_0) = Te^{\frac{1}{i\hbar}\int_{t_0}^{t} dt'\widehat{H}_{\mathrm{QED}}(t')} \tag{4.71}$$

where T denotes the Dyson chronological operator (Sakurai 1985).

For $\widehat{F}(t)$ whose time evolution is solely brought about by the canonical variables in the Heisenberg representation

$$\widehat{F}(t) = \widehat{U}^{\dagger}(t, t_0)\widehat{F}(t_0)\widehat{U}(t, t_0) \tag{4.72}$$

we have

$$i\hbar\frac{\partial}{\partial t}\widehat{F}(t) = \widehat{U}^{\dagger}(t, t_0)\left[\widehat{F}(t_0), \widehat{H}_{\mathrm{QED}}(t)\right]\widehat{U}(t, t_0) = \left[\widehat{F}(t), \widehat{H}_{\mathrm{QED}}^{(H)}(t, t_0)\right] \tag{4.73}$$

where

$$\widehat{H}_{\mathrm{QED}}^{(H)}(t, t_0) = \widehat{U}^{\dagger}(t, t_0)\widehat{H}_{\mathrm{QED}}(t)\widehat{U}(t, t_0) \tag{4.74}$$

with

$$\widehat{H}_{\mathrm{QED}}(t_0) = \widehat{H}_{\mathrm{QED}}^{(H)}(t_0, t_0) \tag{4.75}$$

Alternatively, define

$$\widehat{F}^{(H)}(t, t_0) = \widehat{U}^{\dagger}(t, t_0)\widehat{F}(t)\widehat{U}(t, t_0) \tag{4.76}$$

with

$$\widehat{F}(t_0) = \widehat{F}^{(H)}(t_0, t_0) \tag{4.77}$$

then the time evolution of $\widehat{F}^{(H)}(t, t_0)$ is given as

$$\begin{aligned} i\hbar\frac{\partial}{\partial t}\widehat{F}^{(H)}(t, t_0) &= \widehat{U}^{\dagger}(t, t_0)\left[\widehat{F}(t), \widehat{H}_{\mathrm{QED}}(t)\right]\widehat{U}(t, t_0) \\ &\quad + \widehat{U}^{\dagger}(t, t_0)\left(i\hbar\frac{\partial}{\partial t}\widehat{F}(t)\right)\widehat{U}(t, t_0) \\ &= \left[\widehat{F}^{(H)}(t, t_0), \widehat{H}_{\mathrm{QED}}^{(H)}(t, t_0)\right] \\ &\quad + \widehat{U}^{\dagger}(t, t_0)\left(i\hbar\frac{\partial}{\partial t}\widehat{F}(t)\right)\widehat{U}(t, t_0) \end{aligned} \tag{4.78}$$

The time evolution of $\widehat{H}_{\mathrm{QED}}(t)$ itself should also solely be brought about by the canonical variables in the Heisenberg representation

$$\widehat{H}_{\mathrm{QED}}(t) = \widehat{U}^{\dagger}(t, t_0)\widehat{H}_{\mathrm{QED}}(t_0)\widehat{U}(t, t_0) \tag{4.79}$$

and then we have

$$\begin{aligned} i\hbar\frac{\partial}{\partial t}\widehat{H}_{\mathrm{QED}}(t) &= \widehat{U}^{\dagger}(t, t_0)\left[\widehat{H}_{\mathrm{QED}}(t_0), \widehat{H}_{\mathrm{QED}}(t)\right]\widehat{U}(t, t_0) \\ &= \left[\widehat{H}_{\mathrm{QED}}(t), \widehat{H}_{\mathrm{QED}}^{(H)}(t, t_0)\right] \end{aligned} \tag{4.80}$$

The time evolution of $\widehat{H}_{\mathrm{QED}}^{(H)}(t, t_0)$ is given as

$$i\hbar\frac{\partial}{\partial t}\widehat{H}_{\mathrm{QED}}^{(H)}(t,t_0) = \widehat{U}^{\dagger}(t,t_0)\left(i\hbar\frac{\partial}{\partial t}\widehat{H}_{\mathrm{QED}}(t)\right)\widehat{U}(t,t_0) \tag{4.81}$$

4.3.2.2 Initial Ket Vector with Wave Function

The wave function $\Phi_N(t_0; \omega_1, \ldots, \omega_N, t)$ in the Hilbert space of QED is equipped with the ket vector

$$|\Psi(t_0;t)\rangle_{H \text{ or } S} = \sum_{N=0}^{\infty}\int d\omega_1 \ldots d\omega_N |t_0; \omega_1, \ldots, \omega_N, t\rangle_{H \text{ or } S}\Phi_N(t_0; \omega_1, \ldots, \omega_N, t) \tag{4.82}$$

in terms of the Heisenberg (H) or Schrödinger (S) representation satisfying the Heisenberg equation

$$i\hbar\frac{\partial}{\partial t}|\Psi(t_0;t)\rangle_H = 0 \tag{4.83}$$

or the Schrödinger equation

$$i\hbar\frac{\partial}{\partial t}|\Psi(t_0;t)\rangle_S = \widehat{H}_{\mathrm{QED}}(t)|\Psi(t_0;t)\rangle_S, \quad |\Psi(t_0;t)\rangle_S = \widehat{U}(t,t_0)|\Psi(t_0;t)\rangle_H \tag{4.84}$$

A primitive choice of the basis ket vector may be given as

$$|t_0; \omega_1, \ldots, \omega_N, t\rangle_H = \widehat{U}^{\dagger}(t,t_0)|t_0; \omega_1, \ldots, \omega_N, t_0\rangle_S \tag{4.85}$$

$$|t_0; \omega_1, \ldots, \omega_N, t_0\rangle_H = |t_0; \omega_1, \ldots, \omega_N, t_0\rangle_S \tag{4.86}$$

with

$$\begin{aligned}
&|t_0; \omega_1, \ldots, \omega_N, t_0\rangle_S \\
&= \frac{1}{\sqrt{N_b!}}\widehat{b}^{\dagger}(\omega_{1_b})\cdots\widehat{b}^{\dagger}(\omega_{N_b}) \\
&\times \frac{1}{\sqrt{N_f!}}\widehat{f}^{\dagger}(\omega_{1_f})\cdots\widehat{f}^{\dagger}(\omega_{N_f}) \\
&\times \frac{1}{\sqrt{N_{f^c}!}}\widehat{f}^{c\dagger}\left(\omega_{1_{fc}}\right)\cdots\widehat{f}^{c\dagger}\left(\omega_{N_{fc}}\right)|0\rangle
\end{aligned} \tag{4.87}$$

$$\Phi_N(t_0; \omega_1, \ldots, \omega_N, t) = \Phi_N\left(t_0; \omega_{1_b}, \ldots, \omega_{N_b}, \omega_{1_f}, \ldots, \omega_{N_f}, \omega_{1_{fc}}, \ldots, \omega_{N_{fc}}, t\right) \tag{4.88}$$

$$N = N_b \oplus N_f \oplus N_{f^c} \tag{4.89}$$

$$\omega = \omega_b \otimes \omega_f \otimes \omega_{f^c} \tag{4.90}$$

$$\omega_b, \omega_f, \omega_{f^c} = \{\nu, \vec{p}, \sigma\} \tag{4.91}$$

For permutation P of variables

$$\begin{aligned}&\Phi_N(t_0; \omega_{P1}, \ldots, \omega_{PN}, t)\\&= \Phi_N\left(t_0; \omega_{P_b 1_b}, \ldots, \omega_{P_b N_b}, \omega_{P_f 1_f}, \ldots, \omega_{P_f N_f}, \omega_{P_{fc} 1_{fc}}, \ldots, \omega_{P_{fc} N_{fc}}, t\right)\end{aligned} \tag{4.92}$$

$$P = P_b \otimes P_f \otimes P_{f^c} \tag{4.93}$$

the wave function changes the antisymmetric $(-)$ sign

$$\operatorname{sgn}(P)\Phi_N(t_0; \omega_{P1}, \ldots, \omega_{PN}, t) = \Phi_N(t_0; \omega_1, \ldots, \omega_N, t) \tag{4.94}$$

$$\operatorname{sgn}(P) = (-)^{P_f}(-)^{P_{f^c}} \tag{4.95}$$

Here we have the time-dependent basis ket vector for the Heisenberg (H) representation

$$\begin{aligned}i\hbar\frac{\partial}{\partial t}|t_0; \omega_1, \ldots, \omega_N, t\rangle_H &= -\widehat{U}^\dagger(t, t_0)\widehat{H}_{\text{QED}}(t)|t_0; \omega_1, \ldots, \omega_N, t_0\rangle_S\\&= -\widehat{H}_{\text{QED}}{}^{(H)}(t, t_0)|t_0; \omega_1, \ldots, \omega_N, t\rangle_H\end{aligned} \tag{4.96}$$

and the time-independent one for the Schrödinger (S) representation

$$\begin{aligned}|t_0; \omega_1, \ldots, \omega_N, t\rangle_S &= \widehat{U}(t, t_0)|t_0; \omega_1, \ldots, \omega_N, t\rangle_H\\&= |t_0; \omega_1, \ldots, \omega_N, t_0\rangle_S\end{aligned} \tag{4.97}$$

Using the primitive choice described above, the basis vectors are orthonormal

$$\begin{aligned}&{}_H\langle t_0; \omega_1, \ldots, \omega_N, t|t_0; \omega'_1, \ldots, \omega'_M, t\rangle^H H\\&= {}_S\langle t_0; \omega_1, \ldots, \omega_N, t_0|t_0; \omega'_1, \ldots, \omega'_M, t_0\rangle^S S = \delta_{NM}\delta(\omega - \omega')\\&= \delta_{N_b M_b}\frac{1}{N_b!}\sum_{P_b}\delta_b\left(\omega_{1_b} - \omega'_{P_b 1_b}\right)\cdots\delta_b\left(\omega_{N_b} - \omega'_{P_b N_b}\right)\\&\times \delta_{N_f M_f}\frac{1}{N_f!}\sum_{P_f}(-)^{P_f}\delta_f\left(\omega_{1_f} - \omega'_{P_f 1_f}\right)\cdots\delta_f\left(\omega_{N_f} - \omega'_{P_f N_f}\right)\\&\times \delta_{N_{fc} M_{fc}}\frac{1}{N_{fc}!}\sum_{P_{fc}}(-)^{P_{fc}}\delta_{f^c}\left(\omega_{1_{fc}} - \omega'_{P_{fc} 1_{fc}}\right)\cdots\delta_{f^c}\left(\omega_{N_{fc}} - \omega'_{P_{fc} N_{fc}}\right)\end{aligned} \tag{4.98}$$

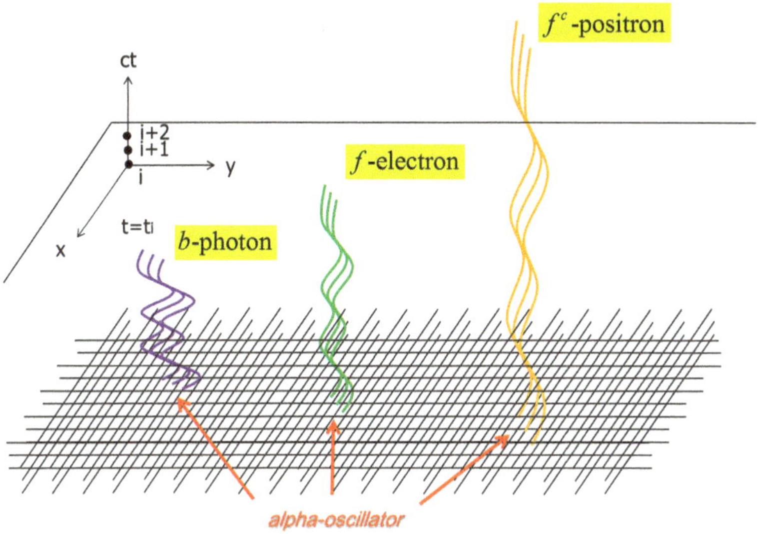

Fig. 4.10 Time evolution of alpha-oscillators

with

$$\delta_b\left(\omega_b-\omega_b'\right)=\delta_{\sigma\sigma'}\delta^3(\vec{p}-\vec{p}')\delta\left(\nu-\nu(|\vec{p}|)_b\right)\delta\left(\nu'-\nu(|\vec{p}'|)_b\right) \tag{4.99}$$

$$\begin{aligned}\delta_f\left(\omega_f-\omega_f'\right)&=\delta_{f^c}\left(\omega_{f^c}-\omega_{f^c}'\right)\\&=\delta_{\sigma\sigma'}\delta^3(\vec{p}-\vec{p}')\delta\left(\nu-\nu(|\vec{p}|)_f\right)\delta\left(\nu'-\nu(|\vec{p}'|)_f\right)\end{aligned} \tag{4.100}$$

Following this primitive choice, we shall find the way how to calculate the time evolution of $\Phi_N(t_0;\omega_1,\ldots,\omega_N,t)$. This will be described in the following Sect. 4.3.2.3 for the case starting at $t_i>t_0$. Taking the limit $t_i\to t_0$, we shall get to the present case.

It should be noted that along with the time evolution of wave functions, we have the time evolution of alpha-oscillators with thermalization among them (see Figs. 4.10 and 4.11).

4.3.2.3 Time Evolution of Ket Vector with Wave Function

The wave function $\Phi_N(\alpha_i,t_i;\omega_1,\ldots,\omega_N,t)$ in the Hilbert space of QED is equipped with the ket vector

$$|\Psi(\alpha_i,t_i;t)\rangle_{H\text{ or }S}=\sum_{N=0}^{\infty}\int d\omega_1\ldots d\omega_N|t_i;\omega_1,\ldots,\omega_N,t\rangle_{H\text{ or }S}\Phi_N(\alpha_i,t_i;\omega_1,\ldots,\omega_N,t) \tag{4.101}$$

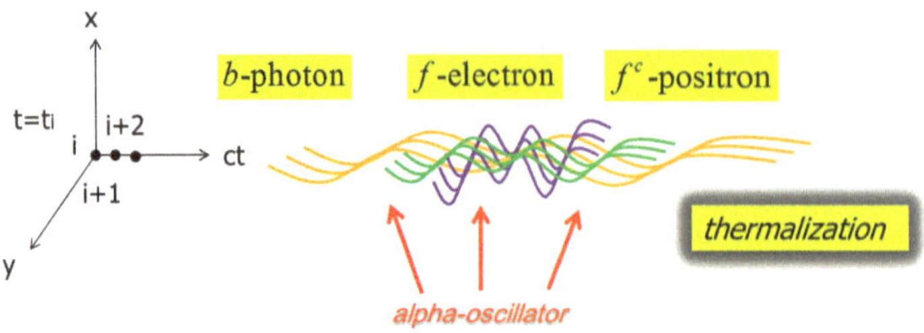

Fig. 4.11 Thermalization of alpha-oscillators

in terms of the Heisenberg (H) or Schrödinger (S) representation satisfying the Heisenberg equation

$$i\hbar\frac{\partial}{\partial t}|\Psi(\alpha_i, t_i; t)\rangle_H = 0 \tag{4.102}$$

or the Schrödinger equation

$$i\hbar\frac{\partial}{\partial t}|\Psi(\alpha_i, t_i; t)\rangle_S = \widehat{H}_{\mathrm{QED}}(t)|\Psi(\alpha_i, t_i; t)\rangle_S \tag{4.103}$$

$$|\Psi(\alpha_i, t_i; t)\rangle_S = \widehat{U}(t, t_i)|\Psi(\alpha_i, t_i; t)\rangle_H \tag{4.104}$$

Here we have used the time evolution operator obeying

$$i\hbar\frac{\partial}{\partial t}\widehat{U}(t, t_i) = \widehat{H}_{\mathrm{QED}}(t)\widehat{U}(t, t_i), \quad i\hbar\frac{\partial}{\partial t}\widehat{U}^\dagger(t, t_i) = -\widehat{U}^\dagger(t, t_i)\widehat{H}_{\mathrm{QED}}(t) \tag{4.105}$$

and the solution

$$\widehat{U}(t, t_i) = Te^{\frac{1}{i\hbar}\int_{t_i}^{t} dt'\widehat{H}_{\mathrm{QED}}(t')} \tag{4.106}$$

Now we find

$$\widehat{F}(t) = \widehat{U}^\dagger(t, t_0)\widehat{F}(t_0)\widehat{U}(t, t_0) = \widehat{U}^\dagger(t, t_i)\widehat{F}(t_i)\widehat{U}(t, t_i) \tag{4.107}$$

and then we have

$$i\hbar\frac{\partial}{\partial t}\widehat{F}(t) = \widehat{U}^\dagger(t, t_i)\left[\widehat{F}(t_i), \widehat{H}_{\mathrm{QED}}(t)\right]\widehat{U}(t, t_i) = \left[\widehat{F}(t), \widehat{H}_{\mathrm{QED}}{}^{(H)}(t, t_i)\right] \tag{4.108}$$

where

$$\widehat{H}_{\mathrm{QED}}^{(H)}(t,t_i) = \widehat{U}^{\dagger}(t,t_i)\widehat{H}_{\mathrm{QED}}(t)\widehat{U}(t,t_i) \tag{4.109}$$

with

$$\widehat{H}_{\mathrm{QED}}^{(H)}(t_i,t_i) = \widehat{H}_{\mathrm{QED}}(t_i) \tag{4.110}$$

and hence

$$i\hbar\frac{\partial}{\partial t}\widehat{H}_{\mathrm{QED}}^{(H)}(t,t_i) = \widehat{U}^{\dagger}(t,t_i)\left(i\hbar\frac{\partial}{\partial t}\widehat{H}_{\mathrm{QED}}(t)\right)\widehat{U}(t,t_i) \tag{4.111}$$

It should be noted that the canonical quantization requires simple rule of the Cauchy data

$$\widehat{F}(t)\Big|_{t=t_i} = \widehat{F}(t_i), \quad i\hbar\frac{\partial}{\partial t}\widehat{F}(t)\Big|_{t=t_i} = \left[\widehat{F}(t_i),\widehat{H}_{\mathrm{QED}}(t_i)\right] \tag{4.112}$$

instead of Eq. (4.108). So the time-dependent nature of $\widehat{H}_{\mathrm{QED}}^{(H)}(t,t_i)$ in the right-hand side of Eq. (4.108) is hidden.

The time evolution of the alternative field operator

$$\widehat{F}^{(H)}(t,t_i) = \widehat{U}^{\dagger}(t,t_i)\widehat{F}(t)\widehat{U}(t,t_i) \tag{4.113}$$

$$\widehat{F}^{(H)}(t_i,t_i) = \widehat{F}(t_i) \tag{4.114}$$

obeys the Heisenberg equation of motion

$$\begin{aligned} i\hbar\frac{\partial}{\partial t}\widehat{F}^{(H)}(t,t_i) &= \widehat{U}^{\dagger}(t,t_i)\left[\widehat{F}(t),\widehat{H}_{\mathrm{QED}}(t)\right]\widehat{U}(t,t_i) + \widehat{U}^{\dagger}(t,t_i)\left(i\hbar\frac{\partial}{\partial t}\widehat{F}(t)\right)\widehat{U}(t,t_i) \\ &= \left[\widehat{F}^{(H)}(t,t_i),\widehat{H}_{\mathrm{QED}}^{(H)}(t,t_i)\right] + \widehat{U}^{\dagger}(t,t_i)\left(i\hbar\frac{\partial}{\partial t}\widehat{F}(t)\right)\widehat{U}(t,t_i) \end{aligned} \tag{4.115}$$

A primitive choice of the basis ket vector may be given as

$$|t_i;\omega_1,\ldots,\omega_N,t\rangle_H = \widehat{U}^{\dagger}(t,t_i)|t_i;\omega_1,\ldots,\omega_N,t_i\rangle_S \tag{4.116}$$

$$|t_i;\omega_1,\ldots,\omega_N,t_i\rangle_H = |t_i;\omega_1,\ldots,\omega_N,t_i\rangle_S \tag{4.117}$$

with

$$
\begin{aligned}
&|t_i;\omega_1,\ldots,\omega_N,t_i\rangle_S \\
&=\widehat{U}^\dagger(t_i,t_0)|t_0;\omega_1,\ldots,\omega_N,t_0\rangle_S \\
&=\widehat{U}^\dagger(t_i,t_0)\frac{1}{\sqrt{N_b!}}\widehat{b}^\dagger(\omega_{1_b})\ldots\widehat{b}^\dagger(\omega_{N_b}) \\
&\quad\times\frac{1}{\sqrt{N_f!}}\widehat{f}^\dagger(\omega_{1_f})\ldots\widehat{f}^\dagger(\omega_{N_f}) \\
&\quad\times\frac{1}{\sqrt{N_{f^c}!}}\widehat{f}^{c\dagger}(\omega_{1_{f^c}})\ldots\widehat{f}^{c\dagger}(\omega_{N_{f^c}})|0\rangle
\end{aligned}
\tag{4.118}
$$

We have the time-dependent basis ket vector for the Heisenberg (H) representation

$$
\begin{aligned}
i\hbar\frac{\partial}{\partial t}|t_i;\omega_1,\ldots,\omega_N,t\rangle_H &= -\widehat{U}^\dagger(t,t_i)\widehat{H}_{\mathrm{QED}}(t)|t_i;\omega_1,\ldots,\omega_N,t_i\rangle_S \\
&= -\widehat{H}_{\mathrm{QED}}^{(H)}(t,t_i)|t_i;\omega_1,\ldots,\omega_N,t\rangle_H
\end{aligned}
\tag{4.119}
$$

and the time-independent one for the Schrödinger (S) representation

$$
\begin{aligned}
|t_i;\omega_1,\ldots,\omega_N,t\rangle_S &= \widehat{U}(t,t_i)|t_i;\omega_1,\ldots,\omega_N,t\rangle_H \\
&= |t_i;\omega_1,\ldots,\omega_N,t_i\rangle_S
\end{aligned}
\tag{4.120}
$$

Using the primitive choice described above, the basis vectors are orthonormal

$$
\begin{aligned}
&{}_H\langle t_i;\omega_1,\ldots,\omega_N,t|t_i;\omega'_1,\ldots,\omega'_M,t\rangle_H \\
&={}_S\langle t_i;\omega_1,\ldots,\omega_N,t_i|t_i;\omega'_1,\ldots,\omega'_M,t_i\rangle_S \\
&={}_S\langle t_0;\omega_1,\ldots,\omega_N,t_0|t_0;\omega'_1,\ldots,\omega'_M,t_0\rangle_S \\
&=\delta_{NM}\delta(\omega-\omega')
\end{aligned}
\tag{4.121}
$$

For an event α_i starting at t_i with $t_0 < t_i;\ \ i = 1,2,3,\ldots$, we set up the initial ket vector for Eq. (4.101) and need to obtain the wave function $\Phi_N(\alpha_i, t_i; \omega_1, \ldots, \omega_N, t)$ satisfying

$$
\begin{aligned}
&i\hbar\frac{\partial}{\partial t}\Phi_N(\alpha_i,t_i;\omega_1,\ldots,\omega_N,t) \\
&=\sum_{M=0}^{\infty}\int d\omega'_1\ldots d\omega'_M H_{NM}(t_i;\omega_1,\ldots,\omega_N,\omega'_1,\ldots,\omega'_M,t)\Phi_M(\alpha_i,t_i;\omega'_1,\ldots,\omega'_M,t)
\end{aligned}
\tag{4.122}
$$

using the time-dependent function

$$\begin{aligned} &H_{NM}\big(t_i;\omega_1,\ldots,\omega_N,\omega_1',\ldots,\omega_M',t\big) \\ &= {}_H\langle t_i;\omega_1,\ldots,\omega_N,t|\widehat{H}_{\mathrm{QED}}{}^{(H)}(t,t_i)\big|t_i;\omega_1',\ldots,\omega_M',t\rangle_H \\ &= {}_S\langle t_i;\omega_1,\ldots,\omega_N,t_i|\widehat{H}_{\mathrm{QED}}(t)\big|t_i;\omega_1',\ldots,\omega_M',t_i\rangle_S \end{aligned} \tag{4.123}$$

$$\frac{\partial}{\partial t}H_{NM}\big(t_i;\omega_1,\ldots,\omega_N,\omega_1',\ldots,\omega_M',t\big)\neq 0 \tag{4.124}$$

Finally, substituting this time-dependent $\Phi_N(\alpha_i,t_i;\omega_1,\ldots,\omega_N,t)$ into Eq. (4.101), we calculate

$$\left\langle \widehat{F}(t)\right\rangle_{\alpha_i,t_i} = \frac{{}_H\langle\Psi(\alpha_i,t_i;t)|\widehat{F}^{(H)}(t,t_i)|\Psi(\alpha_i,t_i;t)\rangle_H}{{}_H\langle\Psi(\alpha_i,t_i;t)|\Psi(\alpha_i,t_i;t)\rangle_H} \tag{4.125}$$

for each event α_i starting at t_i with $t_0<t_i<t$; $i=1,2,3,\ldots$ developing onward with $x^\mu=(ct,\vec{x})$ at position $\vec{x}$ with time t using

$$\begin{aligned} &{}_H\langle\Psi(\alpha_i,t_i;t)|\widehat{F}^{(H)}(t,t_i)|\Psi(\alpha_i,t_i;t)\rangle_H \\ &= \sum_{N=0}^{\infty}\sum_{M=0}^{\infty}\int d\omega_1\ldots d\omega_N\int d\omega_1'\cdots d\omega_M' \\ &\quad\times {}_H\langle t_i;\omega_1,\ldots,\omega_N,t|\widehat{F}^{(H)}(t,t_i)\big|t_i;\omega_1',\ldots,\omega_M',t\rangle_H \\ &\quad\times \Phi_N{}^*(\alpha_i,t_i;\omega_1,\ldots,\omega_N,t)\Phi_M\big(\alpha_i,t_i;\omega_1',\ldots,\omega_M',t\big) \\ &= \sum_{N=0}^{\infty}\sum_{M=0}^{\infty}\int d\omega_1\ldots d\omega_N\int d\omega_1'\ldots d\omega_M' \\ &\quad\times {}_S\langle t_i;\omega_1,\ldots,\omega_N,t_i|\widehat{F}(t)\big|t_i;\omega_1',\ldots,\omega_M',t_i\rangle_S \\ &\quad\times \Phi_N{}^*(\alpha_i,t_i;\omega_1,\ldots,\omega_N,t)\Phi_M\big(\alpha_i,t_i;\omega_1',\ldots,\omega_M',t\big) \end{aligned} \tag{4.126}$$

$$\begin{aligned} {}_H\langle\Psi(\alpha_i,t_i;t)|\Psi(\alpha_i,t_i;t)\rangle_H &= \sum_{N=0}^{\infty}\int d\omega_1\ldots d\omega_N|\Phi_N(\alpha_i,t_i;\omega_1,\ldots,\omega_N,t)|^2 \\ &= \sum_{N=0}^{\infty}\int d\omega_1\ldots d\omega_N|\Phi_N(\alpha_i,t_i;\omega_1,\ldots,\omega_N,t_i)|^2 \end{aligned} \tag{4.127}$$

4.4 Normal Mode

4.4.1 Particle Picture

Since the Hamiltonian is time dependent, we need time-dependent renormalization for particle picture (Tachibana 2016). The coarse-grained collective picture of particle is realized through three steps with the time-dependent renormalization (see Fig. 4.12).

They are (I) particle spectrum condition, the dispersion rule of the frequency ν $(|\vec{p}|)$ over and above the alpha-resonance condition Eq. (4.69); (II) algebra normal mode condition, the coarse-grained collective picture of particle over and above the sub-particle alpha-oscillators; and (III) field operator renormalization condition, the renormalized field operators for the resultant particles. Since the Hamiltonian $\widehat{H}_{\mathrm{QED}}$ (t) is time dependent, so are the steps (II) and (III). The renormalization constants $\widehat{z}(t)$ are therefore time-dependent q-numbers, serving to collect sub-particle alpha-oscillators among the time-dependent $\widehat{H}_{\mathrm{QED}}(t)$. The concrete procedures are described as follows. It should be noted that the particle is the complementary picture of the field normal mode (see Fig. 4.13).

4.4.2 Electromagnetic Field Renormalization

(I) Particle spectrum condition

$$c{p_{\mathrm{photon}}}^0 = h\nu_{\mathrm{photon}} = c|\vec{p}| \tag{4.128}$$

is the dispersion rule of photon, and consequently the alpha-resonance condition Eq. (4.69) is further restricted to

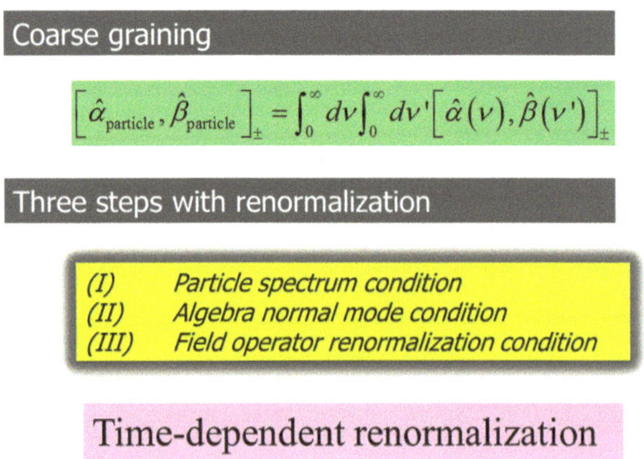

Fig. 4.12 Time-dependent renormalization with coarse-grained collective picture of particle

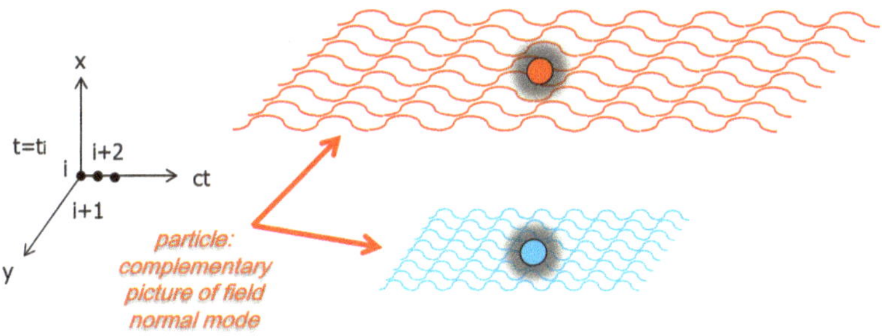

Fig. 4.13 Normal mode with particle

$$cp^0\left(\nu(|\vec{p}|)_b, |\vec{p}|\right) = h\nu(|\vec{p}|)_b = c|\vec{p}| \quad \left(i.e., = cp^0{}_{\text{photon}} = h\nu_{\text{photon}}\right) \tag{4.129}$$

(II) Algebra normal mode condition

$$\left[\frac{1}{\sqrt{\widehat{z}_{b(\widetilde{\omega})}(t)}}\widehat{b}_{\text{photon}}(\widetilde{\omega}(t)), \widehat{H}_{\text{QED}}(t)\right] = h\nu_{\text{photon}}(t)\frac{1}{\sqrt{\widehat{z}_{b(\widetilde{\omega})}(t)}}\widehat{b}_{\text{photon}}(\widetilde{\omega}(t)) \tag{4.130}$$

with the coarse-grained commutation relationship

$$\begin{aligned}\left[\widehat{b}_{\text{photon}}(\widetilde{\omega}(t)), \widehat{b}^{\dagger}{}_{\text{photon}}(\widetilde{\omega}'(t))\right] &= \int_0^{\infty} d\nu \int_0^{\infty} d\nu' \left[\widehat{b}(\omega), \widehat{b}^{\dagger}(\omega')\right] \\ &= \delta(\widetilde{\omega}(t) - \widetilde{\omega}'(t)) = \delta_{\sigma\sigma'}\delta^3(\vec{p} - \vec{q})\end{aligned} \tag{4.131}$$

where

$$\widetilde{\omega}(t) = \int_0^{\infty} d\nu\omega\delta\left(\nu - \nu_{\text{photon}}(t)\right) = \left\{\nu_{\text{photon}}(t), \vec{p}, \sigma\right\} \tag{4.132}$$

(III) Field operator renormalization condition

$$\begin{aligned}\frac{1}{\sqrt{\widehat{z}_b(t)}}\hat{\vec{b}}_{\text{photon}}(x) = &\frac{\sqrt{4\pi\hbar^2 c}}{\sqrt{(2\pi\hbar)^3}}\sum_{\sigma=\pm1}\int_{-\infty}^{\infty}\frac{d^3\vec{p}}{\sqrt{2p_{\text{photon}}{}^0\left(\nu_{\text{photon}}(t), |\vec{p}|\right)}} \\ &\times \frac{1}{\sqrt{\widehat{z}_{b(\widetilde{\omega})}(t)}}\widehat{b}_{\text{photon}}(\widetilde{\omega}(t))e^{-i2\pi\nu_{\text{photon}}(t)t}\vec{e}(\vec{p}, \sigma)e^{i\vec{x}\bullet\vec{p}/\hbar}\end{aligned} \tag{4.133}$$

using

$$\hat{\vec{b}}(x) = \frac{1}{\sqrt{\widehat{z}_b(t)}} \hat{\vec{b}}_{\text{photon}}(x) \tag{4.134}$$

$$\widehat{b}(\omega) = \frac{1}{\sqrt{\widehat{z}_{b(\widetilde{\omega})}(t)}} \widehat{b}_{\text{photon}}(\widetilde{\omega}(t))\delta(\nu - \nu_{\text{photon}}(t)) \tag{4.135}$$

4.4.3 *The Dirac Field Renormalization*

(I) Particle spectrum condition

$$cp^0{}_{\text{electron}} = h\nu_{\text{electron}} = c\sqrt{(mc)^2 + |\vec{p}|^2} \tag{4.136}$$

is the dispersion rule of the Dirac particle, and consequently the alpha-resonance condition Eq. (4.69) is further restricted to

$$\begin{aligned} cp^0\left(\nu(|\vec{p}|)_f, |\vec{p}|\right) &= h\nu(|\vec{p}|)_f \\ &= c\sqrt{(mc)^2 + |\vec{p}|^2} \;\; \left(i.e., = cp^0{}_{\text{electron}} = h\nu_{\text{electron}}\right) \end{aligned} \tag{4.137}$$

(II) Algebra normal mode condition

$$\left[\frac{1}{\sqrt{\widehat{z}_{f(\widetilde{\omega})}(t)}} \widehat{f}_{\text{electron}}(\widetilde{\omega}(t)), \widehat{H}_{\text{QED}}(t)\right] = h\nu_{\text{electron}}(t) \frac{1}{\sqrt{\widehat{z}_{f(\widetilde{\omega})}(t)}} \widehat{f}_{\text{electron}}(\widetilde{\omega}(t)) \tag{4.138}$$

$$\begin{aligned} &\left[\frac{1}{\sqrt{\widehat{z}_{f^c(\widetilde{\omega})}(t)}} \widehat{f}_{\text{positron}}{}^c(\widetilde{\omega}(t)), \widehat{H}_{\text{QED}}(t)\right] \\ &= h\nu_{\text{electron}}(t) \frac{1}{\sqrt{\widehat{z}_{f^c(\widetilde{\omega})}(t)}} \widehat{f}_{\text{positron}}{}^c(\widetilde{\omega}(t)) \end{aligned} \tag{4.139}$$

with the coarse-grained commutation relationship

$$\left\{\widehat{f}_{\text{electron}}(\widetilde{\omega}(t)),\widehat{f}^{\dagger}{}_{\text{electron}}(\widetilde{\omega}'(t))\right\} = \int_0^\infty d\nu \int_0^\infty d\nu' \left\{\widehat{f}(\omega),\widehat{f}^{\dagger}(\omega')\right\} \\ = \delta(\widetilde{\omega}(t) - \widetilde{\omega}'(t)) = \delta_{\sigma\sigma'}\delta^3(\vec{p} - \vec{q}) \tag{4.140}$$

$$\left\{\widehat{f}_{\text{positron}}{}^{c}(\widetilde{\omega}(t)),\widehat{f}_{\text{positron}}{}^{c\dagger}(\widetilde{\omega}'(t))\right\} = \int_0^\infty d\nu \int_0^\infty d\nu' \left\{\widehat{f}^{c}(\omega),\widehat{f}^{c\dagger}(\omega')\right\} \\ = \delta(\widetilde{\omega}(t) - \widetilde{\omega}'(t)) = \delta_{\sigma\sigma'}\delta^3(\vec{p} - \vec{q}) \tag{4.141}$$

where

$$\widetilde{\omega}(t) = \int_0^\infty d\nu\omega\delta(\nu - \nu_{\text{electron}}(t)) = \{\nu_{\text{electron}}(t), \vec{p}, \sigma\} \tag{4.142}$$

(III) Field operator renormalization condition

$$\frac{1}{\sqrt{\widehat{z}_f(t)}}\widehat{f}_{\text{electron}}(x) = \frac{1}{\sqrt{(2\pi\hbar)^3}}\sum_{\sigma=\pm\frac{1}{2}}\int_{-\infty}^{\infty} d^3\vec{p} \\ \times \frac{1}{\sqrt{\widehat{z}_{f(\widetilde{\omega})}(t)}}\widehat{f}_{\text{electron}}(\widetilde{\omega}(t))e^{-i2\pi\nu_{\text{electron}}(t)t}u(\vec{p},\sigma)e^{i\vec{x}\cdot\vec{p}/\hbar} \tag{4.143}$$

$$\frac{1}{\sqrt{\widehat{z}_{f^c}(t)}}\widehat{f}_{\text{positron}}{}^{c}(x) = \frac{1}{\sqrt{(2\pi\hbar)^3}}\sum_{\sigma=\pm\frac{1}{2}}\int_{-\infty}^{\infty} d^3\vec{p} \\ \times \frac{1}{\sqrt{\widehat{z}_{f^c(\widetilde{\omega})}(t)}}\widehat{f}_{\text{positron}}{}^{c}(\widetilde{\omega}(t))e^{-i2\pi\nu_{\text{electron}}(t)t}v^{\dagger}(\vec{p},\sigma)e^{i\vec{x}\cdot\vec{p}/\hbar} \tag{4.144}$$

using

$$\widehat{f}(x) = \frac{1}{\sqrt{\widehat{z}_f(t)}}\widehat{f}_{\text{electron}}(x) \tag{4.145}$$

$$\widehat{f}(\omega) = \frac{1}{\sqrt{\widehat{z}_{f(\widetilde{\omega})}(t)}}\widehat{f}_{\text{electron}}(\widetilde{\omega}(t))\delta(\nu - \nu_{\text{electron}}(t)) \tag{4.146}$$

$$\widehat{f}^{c}(x) = \frac{1}{\sqrt{\widehat{z}_{f^c}(t)}}\widehat{f}_{\text{positron}}{}^{c}(x) \tag{4.147}$$

$$\widehat{f}^c(\omega) = \frac{1}{\sqrt{\widehat{z}_{f^c(\widetilde{\omega})}(t)}} \widehat{f}_{\text{positron}}{}^c(\widetilde{\omega}(t))\delta(\nu - \nu_{\text{electron}}(t)) \tag{4.148}$$

4.4.4 Renormalized Ket Vector and Wave Function

Let the dual Cauchy problem in QED for an event α_i starting at t_i with $t_0 < t_i$; $i = 1, 2, 3, \ldots$ be described by the renormalized wave function $\widetilde{\Phi}_N(\alpha_i, t_i; \widetilde{\omega}_1(t_i), \ldots, \widetilde{\omega}_N(t_i), t)$ in the Hilbert space of QED equipped with the renormalized ket vector (see Fig. 4.14)

$$\left|\widetilde{\Psi}(\alpha_i, t_i; t)\right\rangle_{H \text{ or } S} = \sum_{N=0}^{\infty} \int d\widetilde{\omega}_1(t_i) \ldots d\widetilde{\omega}_N(t_i) |t_i; \widetilde{\omega}_1(t_i), \ldots, \widetilde{\omega}_N(t_i), t\rangle_{H \text{ or } S} \widetilde{\Phi}_N(\alpha_i, t_i; \widetilde{\omega}_1(t_i), \ldots, \widetilde{\omega}_N(t_i), t) \tag{4.149}$$

in terms of the Heisenberg (H) or Schrödinger (S) representation satisfying the Heisenberg equation

$$i\hbar \frac{\partial}{\partial t} \left|\widetilde{\Psi}(\alpha_i, t_i; t)\right\rangle_H = 0 \tag{4.150}$$

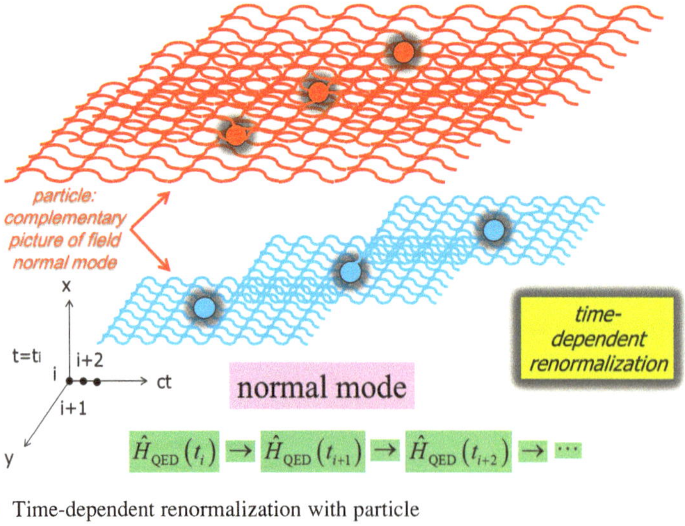

Fig. 4.14 Time-dependent renormalization with particle

or the Schrödinger equation

$$i\hbar\frac{\partial}{\partial t}\left|\widetilde{\Psi}(\alpha_i,t_i;t)\right\rangle_S=\widehat{H}_{\mathrm{QED}}(t)\left|\widetilde{\Psi}(\alpha_i,t_i;t)\right\rangle_S,\quad \left|\widetilde{\Psi}(\alpha_i,t_i;t)\right\rangle_S = \widehat{U}(t,t_i)\left|\widetilde{\Psi}(\alpha_i,t_i;t)\right\rangle_H \tag{4.151}$$

The basis ket vectors are given by

$$|t_i;\widetilde{\omega}_1(t_i),\ldots,\widetilde{\omega}_N(t_i),t\rangle_H=\widehat{U}^\dagger(t,t_i)|t_i;\widetilde{\omega}_1(t_i),\ldots,\widetilde{\omega}_N(t_i),t_i\rangle_S \tag{4.152}$$

$$|t_i;\widetilde{\omega}_1(t_i),\ldots,\widetilde{\omega}_N(t_i),t_i\rangle_H=|t_i;\widetilde{\omega}_1(t_i),\ldots,\widetilde{\omega}_N(t_i),t_i\rangle_S \tag{4.153}$$

with

$$\begin{aligned}
&|t_i;\widetilde{\omega}_1(t_i),\ldots,\widetilde{\omega}_N(t_i),t_i\rangle_S\\
&=\frac{1}{\sqrt{N_b!}}\left(\frac{1}{\sqrt{\widehat{z}_{b(\widetilde{\omega}_{1_b})}(t_i)}}\widehat{b}_{\mathrm{photon}}(\widetilde{\omega}_{1_b}(t_i))\right)^\dagger\cdots\left(\frac{1}{\sqrt{\widehat{z}_{b(\widetilde{\omega}_{N_b})}(t_i)}}\widehat{b}_{\mathrm{photon}}(\widetilde{\omega}_{N_b}(t_i))\right)^\dagger\\
&\times\frac{1}{\sqrt{N_f!}}\left(\frac{1}{\sqrt{\widehat{z}_{f(\widetilde{\omega}_{1_f})}(t_i)}}\widehat{f}_{\mathrm{electron}}(\widetilde{\omega}_{1_f}(t_i))\right)^\dagger\cdots\left(\frac{1}{\sqrt{\widehat{z}_{f(\widetilde{\omega}_{N_f})}(t_i)}}\widehat{f}_{\mathrm{electron}}(\widetilde{\omega}_{N_f}(t_i))\right)^\dagger\\
&\times\frac{1}{\sqrt{N_{f^c}!}}\left(\frac{1}{\sqrt{\widehat{z}_{f^c\left(\widetilde{\omega}_{1_{f^c}}\right)}(t_i)}}\widehat{f}_{\mathrm{positron}}{}^c\left(\widetilde{\omega}_{1_{f^c}}(t_i)\right)\right)^\dagger\cdots\left(\frac{1}{\sqrt{\widehat{z}_{f^c\left(\widetilde{\omega}_{N_{f^c}}\right)}(t_i)}}\widehat{f}_{\mathrm{positron}}{}^c\left(\widetilde{\omega}_{N_{f^c}}(t_i)\right)\right)^\dagger|0\rangle
\end{aligned} \tag{4.154}$$

We have the time-dependent basis ket vector

$$\begin{aligned}
i\hbar\frac{\partial}{\partial t}|t_i;\widetilde{\omega}_1(t_i),\ldots,\widetilde{\omega}_N(t_i),t\rangle_H&=-\widehat{H}_{\mathrm{QED}}{}^{(H)}(t,t_i)|t_i;\widetilde{\omega}_1(t_i),\ldots,\widetilde{\omega}_N(t_i),t\rangle_H\\
&=-\widehat{U}^\dagger(t,t_i)\widehat{H}_{\mathrm{QED}}(t)|t_i;\widetilde{\omega}_1(t_i),\ldots,\widetilde{\omega}_N(t_i),t_i\rangle_S
\end{aligned} \tag{4.155}$$

and the time-independent one

$$\begin{aligned}
&|t_i;\widetilde{\omega}_1(t_i),\ldots,\widetilde{\omega}_N(t_i),t\rangle_S\\
&=\widehat{U}(t,t_i)|t_i;\widetilde{\omega}_1(t_i),\ldots,\widetilde{\omega}_N(t_i),t\rangle_H\\
&=|t_i;\widetilde{\omega}_1(t_i),\ldots,\widetilde{\omega}_N(t_i),t_i\rangle_S
\end{aligned} \tag{4.156}$$

Using the primitive choice described above, the basis vectors are orthonormal

$$
\begin{aligned}
&{}_H\langle t_i;\widetilde{\omega}_1(t_i),\ldots,\widetilde{\omega}_N(t_i),t|t_i;\widetilde{\omega}'_1(t_i),\ldots,\widetilde{\omega}'_M(t_i),t\rangle_H\\
&={}_S\langle t_i;\widetilde{\omega}_1(t_i),\ldots,\widetilde{\omega}_N(t_i),t_i|t_i;\widetilde{\omega}'_1(t_i),\ldots,\widetilde{\omega}'_M(t_i),t_i\rangle_S\\
&=\delta_{NM}\delta(\widetilde{\omega}(t_i)-\widetilde{\omega}'(t_i))
\end{aligned}
\tag{4.157}
$$

4.4.5 Formal Solutions of $\widehat{z}_{\alpha(\widetilde{\omega})}(t)$

Step (III) in Sect. 4.4.1 utilizes the q-number renormalization constant in Step (II), whose solution is demonstrated here.

4.4.5.1 Setup

In Step (II) in Sect. 4.4.1, the algebra normal mode conditions in Eqs. (4.129), (4.138), and (4.139) read

$$
\left[\frac{1}{\sqrt{\widehat{z}_{\alpha(\widetilde{\omega})}(t)}}\widehat{\alpha}_{\text{particle}}(\widetilde{\omega}(t)),\widehat{H}_{\text{QED}}(t)\right]=h\nu_{\text{particle}}(t)\frac{1}{\sqrt{\widehat{z}_{\alpha(\widetilde{\omega})}(t)}}\widehat{\alpha}_{\text{particle}}(\widetilde{\omega}(t))
\tag{4.158}
$$

where $\widehat{z}_{\alpha(\widetilde{\omega})}(t)$ denotes the q-number renormalization constant of $\widehat{\alpha}_{\text{particle}}(\widetilde{\omega}(t))$ with the energy $h\nu_{\text{particle}}(t)$. Using the operator identity

$$
\left[\widehat{A}\widehat{B},\widehat{C}\right]=\widehat{A}\left[\widehat{B},\widehat{C}\right]+\left[\widehat{A},\widehat{C}\right]\widehat{B}
\tag{4.159}
$$

the left-hand side of Eq. (4.158) is reduced to

$$
\begin{aligned}
&\left[\frac{1}{\sqrt{\widehat{z}_{\alpha(\widetilde{\omega})}(t)}}\widehat{\alpha}_{\text{particle}}(\widetilde{\omega}(t)),\widehat{H}_{\text{QED}}(t)\right]\\
&=\frac{1}{\sqrt{\widehat{z}_{\alpha(\widetilde{\omega})}(t)}}\left[\widehat{\alpha}_{\text{particle}}(\widetilde{\omega}(t)),\widehat{H}_{\text{QED}}(t)\right]+\left[\frac{1}{\sqrt{\widehat{z}_{\alpha(\widetilde{\omega})}(t)}},\widehat{H}_{\text{QED}}(t)\right]\widehat{\alpha}_{\text{particle}}(\widetilde{\omega}(t))
\end{aligned}
\tag{4.160}
$$

Using Eqs. (4.158) and (4.160), we have

$$\sqrt{\widehat{z}_{\alpha(\widetilde{\omega})}(t)}\left[\frac{1}{\sqrt{\widehat{z}_{\alpha(\widetilde{\omega})}(t)}}, \widehat{H}_{\mathrm{QED}}(t)\right]\widehat{\alpha}_{\mathrm{particle}}(\widetilde{\omega}(t)) = h\nu_{\mathrm{particle}}(t)\widehat{\alpha}_{\mathrm{particle}}(\widetilde{\omega}(t)) - \left[\widehat{\alpha}_{\mathrm{particle}}(\widetilde{\omega}(t)), \widehat{H}_{\mathrm{QED}}(t)\right] \tag{4.161}$$

If $\widehat{H}_{\mathrm{QED}}(t)$ were the time-independent free field Hamiltonian $\widehat{H}_{\text{free field}}$, then the right-hand side of Eq. (4.161) were zero because of the consequent time-independent free particle operator $\widehat{\alpha}_{\text{free particle}}$ with the energy $h\nu_{\text{free particle}}$; and then the left-hand side of Eq. (4.161) were zero, since the q-number renormalization constant should then be identity. Thus, the right-hand side of Eq. (4.161) is not null for $\widehat{H}_{\mathrm{QED}}(t)$ and then we may write

$$h\nu_{\mathrm{particle}}(t)\widehat{\alpha}_{\mathrm{particle}}(\widetilde{\omega}(t)) - \left[\widehat{\alpha}_{\mathrm{particle}}(\widetilde{\omega}(t)), \widehat{H}_{\mathrm{QED}}(t)\right] = \sqrt{\widehat{z}_{\alpha(\widetilde{\omega})}(t)}\widehat{C}_{\alpha(\widetilde{\omega})}(t)\widehat{\alpha}_{\mathrm{particle}}(\widetilde{\omega}(t)) \tag{4.162}$$

with the non-null $\widehat{C}_{\alpha(\widetilde{\omega})}(t)$ so defined in the right-hand side of Eq. (4.162)

$$\widehat{C}_{\alpha(\widetilde{\omega})}(t) \neq 0 \tag{4.163}$$

Using Eqs. (4.160) and (4.162), we arrive at

$$\left[\frac{1}{\sqrt{\widehat{z}_{\alpha(\widetilde{\omega})}(t)}}, \widehat{H}_{\mathrm{QED}}(t)\right] = \widehat{C}_{\alpha(\widetilde{\omega})}(t) \tag{4.164}$$

This is just an operator equation Eq. (4.166) in Sect. 4.4.5.2. With the proof Eq. (4.168) given, a solution may be found using Eq. (4.167) as

$$\frac{1}{\sqrt{\widehat{z}_{\alpha(\widetilde{\omega})}(t)}} = f\left(\widehat{B}\right) + \sum_{n=0}^{\infty}\widehat{H}_{\mathrm{QED}}(t)^{n}\widehat{C}_{\alpha(\widetilde{\omega})}(t)\left(\frac{1}{\widehat{H}_{\mathrm{QED}}(t)}\right)^{n+1}, \quad \left[\widehat{H}_{\mathrm{QED}}(t), f\left(\widehat{B}\right)\right] = 0, \quad \left[\widehat{H}_{\mathrm{QED}}(t), \widehat{B}\right] = 0 \tag{4.165}$$

modulo a function f of $\widehat{B}$ commutable with $\widehat{H}_{\mathrm{QED}}(t)$.

4.4.5.2 Solutions

Let us solve $\widehat{X}$ in an operator equation

$$\left[\widehat{X},\widehat{A}\right]=\widehat{C},\quad \widehat{A}\neq 0 \tag{4.166}$$

A solution may be found as

$$\widehat{X}=f\left(\widehat{B}\right)+\sum_{n=0}^{\infty}\widehat{A}^{n}\widehat{C}\left(\frac{1}{\widehat{A}}\right)^{n+1},\quad \left[\widehat{A},f\left(\widehat{B}\right)\right]=0,\quad \left[\widehat{A},\widehat{B}\right]=0 \tag{4.167}$$

modulo a function f of $\widehat{B}$ commutable with $\widehat{A}$.

Proof

$$\begin{aligned}\left[\widehat{X},\widehat{A}\right]&=\left[f\left(\widehat{B}\right)+\sum_{n=0}^{\infty}\widehat{A}^{n}\widehat{C}\left(\frac{1}{\widehat{A}}\right)^{n+1},\widehat{A}\right]=\sum_{n=0}^{\infty}\left[\widehat{A}^{n}\widehat{C}\left(\frac{1}{\widehat{A}}\right)^{n+1},\widehat{A}\right]\\&=\sum_{n=0}^{\infty}\left(\widehat{A}^{n}\widehat{C}\left(\frac{1}{\widehat{A}}\right)^{n}-\widehat{A}^{n+1}\widehat{C}\left(\frac{1}{\widehat{A}}\right)^{n+1}\right)=\widehat{A}^{0}\widehat{C}\left(\frac{1}{\widehat{A}}\right)^{0}=\widehat{C}\end{aligned} \tag{4.168}$$

4.4.5.3 Example

Let us find an example of $\widehat{X}$ in an operator Eq. (4.166) whose solution may be found as Eq. (4.167) modulo a function f of $\widehat{B}$ commutable with $\widehat{A}$.

The Pauli matrices

$$\sigma_x=\begin{pmatrix}0&1\\1&0\end{pmatrix},\sigma_y=\begin{pmatrix}0&-i\\i&0\end{pmatrix},\sigma_z=\begin{pmatrix}1&0\\0&-1\end{pmatrix} \tag{4.169}$$

satisfy

$$\begin{aligned}(\sigma_x)^2=\left(\sigma_y\right)^2=(\sigma_z)^2=1\ \text{means}\ (\sigma_x)^{-1}=\sigma_x,\left(\sigma_y\right)^{-1}=\sigma_y,(\sigma_z)^{-1}\\=\sigma_z\end{aligned} \tag{4.170}$$

$$\sigma_z\sigma_y=-i\sigma_x,\sigma_y\sigma_z=i\sigma_x \tag{4.171}$$

$$\sigma_x\sigma_z=-i\sigma_y,\sigma_z\sigma_x=i\sigma_y \tag{4.172}$$

$$\sigma_y\sigma_x=-i\sigma_z,\sigma_x\sigma_y=i\sigma_z \tag{4.173}$$

Let X in

$$X\sigma_y - \sigma_y X = 2i\sigma_z \tag{4.174}$$

be found as

$$X = f(B) + \sum_{n=0}^{\infty} (\sigma_y)^n 2i\sigma_z (\sigma_y)^{-(n+1)}, \quad [\sigma_y, f(B)] = 0, \quad [\sigma_y, B] = 0 \tag{4.175}$$

Proof

$$\begin{aligned}
\sum_{n=0}^{\infty}(\sigma_y)^n 2i\sigma_z(\sigma_y)^{-(n+1)} &= \sum_{n=0}^{\infty}(\sigma_y)^n 2i\sigma_z(\sigma_y)^{n+1} \\
&= 2i(\sigma_z\sigma_y + \sigma_y\sigma_z(\sigma_y)^2 + (\sigma_y)^2\sigma_z(\sigma_y)^{2+1} + (\sigma_y)^{2+1}\sigma_z(\sigma_y)^{2+2} + \cdots) \\
&= 2i(\sigma_z\sigma_y + \sigma_y\sigma_z + \sigma_z\sigma_y + \sigma_y\sigma_z + \cdots) \\
&= 2i(-i\sigma_x + i\sigma_x - i\sigma_x + i\sigma_x - \cdots) \\
&= 2i(-i\sigma_x)(1 - 1 + 1 - 1 + \cdots) \\
&= 2i(-i\sigma_x)\frac{1}{2} \\
&= \sigma_x
\end{aligned} \tag{4.176}$$

where we have used Eqs. (4.170) and (4.171) and the Cesaro sum

$$1 - 1 + 1 - 1 + \cdots = \frac{1}{2} \tag{4.177}$$

It follows that Eq. (4.174) is solved by using Eq. (4.175) as

$$\begin{aligned}
&X\sigma_y - \sigma_y X \\
&= \left(f(B) + \sum_{n=0}^{\infty}(\sigma_y)^n 2i\sigma_z(\sigma_y)^{-(n+1)}\right)\sigma_y - \sigma_y\left(f(B) + \sum_{n=0}^{\infty}(\sigma_y)^n 2i\sigma_z(\sigma_y)^{-(n+1)}\right) \\
&= \left(\sum_{n=0}^{\infty}(\sigma_y)^n 2i\sigma_z(\sigma_y)^{-(n+1)}\right)\sigma_y - \sigma_y\left(\sum_{n=0}^{\infty}(\sigma_y)^n 2i\sigma_z(\sigma_y)^{-(n+1)}\right) \\
&= \sigma_x\sigma_y - \sigma_y\sigma_x \\
&= 2i\sigma_z
\end{aligned} \tag{4.178}$$

where we have used Eqs. (4.176) and (4.173).

4.5 Discussions

The commutation relationships of particles are obtained by coarse graining of the alpha-oscillators. The generic feature may be written as

$$\left[\widehat{\alpha}_{\text{particle}}, \widehat{\beta}_{\text{particle}}\right]_{\pm} = \int_0^{\infty} d\nu \int_0^{\infty} d\nu' \left[\widehat{\alpha}(\nu), \widehat{\beta}(\nu')\right]_{\pm} \tag{4.179}$$

The coarse graining is also applied to the wave function equipped with the ket vector of the renormalized particles.

For an event α_i starting at t_i with $t_0 < t_i$; $i = 1, 2, 3, \ldots$, we set up the initial ket vector for Eq. (4.149) and need to obtain the wave function $\widetilde{\Phi}_N(\alpha_i, t_i; \widetilde{\omega}_1(t_i), \ldots, \widetilde{\omega}_N(t_i), t)$ satisfying

$$\begin{aligned} & i\hbar\frac{\partial}{\partial t}\widetilde{\Phi}_N(\alpha_i, t_i; \widetilde{\omega}_1(t_i), \ldots, \widetilde{\omega}_N(t_i), t) \\ & = \sum_{M=0}^{\infty}\int d\widetilde{\omega}'_1(t_i)\ldots d\widetilde{\omega}'_M(t_i) \\ & \times H_{NM}\left(t_i; \widetilde{\omega}_1(t_i), \ldots, \widetilde{\omega}_N(t_i), \widetilde{\omega}'_1(t_i), \ldots, \widetilde{\omega}'_M(t_i), t\right)\widetilde{\Phi}_M\left(\alpha_i, t_i; \widetilde{\omega}'_1(t_i), \ldots, \widetilde{\omega}'_M(t_i), t\right) \end{aligned} \tag{4.180}$$

using the time-dependent function (see Fig. 4.15)

$$\begin{aligned} & H_{NM}\left(t_i; \widetilde{\omega}_1(t_i), \ldots, \widetilde{\omega}_N(t_i), \widetilde{\omega}'_1(t_i), \ldots, \widetilde{\omega}'_M(t_i), t\right) \\ & = {}_H\langle t_i; \widetilde{\omega}_1(t_i), \ldots, \widetilde{\omega}_N(t_i), t|\widehat{H}_{\text{QED}}{}^{(H)}(t, t_i)|t_i; \widetilde{\omega}'_1(t_i), \ldots, \widetilde{\omega}'_M(t_i), t\rangle_H \\ & = {}_S\langle t_i; \widetilde{\omega}_1(t_i), \ldots, \widetilde{\omega}_N(t_i), t_i|\widehat{H}_{\text{QED}}(t)|t_i; \widetilde{\omega}'_1(t_i), \ldots, \widetilde{\omega}'_M(t_i), t_i\rangle_S \end{aligned} \tag{4.181}$$

$$\frac{\partial}{\partial t}H_{NM}\left(t_i; \widetilde{\omega}_1(t_i), \ldots, \widetilde{\omega}_N(t_i), \widetilde{\omega}'_1(t_i), \ldots, \widetilde{\omega}'_M(t_i), t\right) \neq 0 \tag{4.182}$$

Substituting this time-dependent $\widetilde{\Phi}_N(\alpha_i, t_i; \widetilde{\omega}_1(t_i), \ldots, \widetilde{\omega}_N(t_i), t)$ into Eq. (4.149), we calculate the expectation value

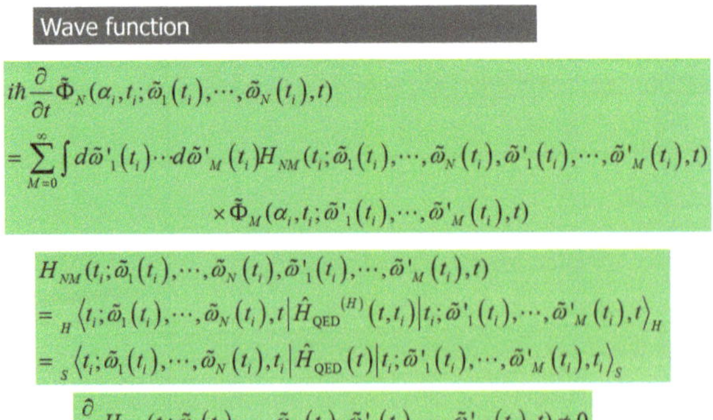

Fig. 4.15 Time evolution of the renormalized wave function

$$\left\langle \widetilde{\widehat{F}}(t) \right\rangle_{\alpha_i, t_i} = \frac{{}_H\left\langle \widetilde{\Psi}(\alpha_i, t_i; t) \left| \widehat{F}^{(H)}(t, t_i) \right| \widetilde{\Psi}(\alpha_i, t_i; t) \right\rangle_H}{{}_H\left\langle \widetilde{\Psi}(\alpha_i, t_i; t) | \widetilde{\Psi}(\alpha_i, t_i; t) \right\rangle_H} \tag{4.183}$$

for each event α_i starting at t_i with $t_0 < t_i < t$; $i = 1, 2, 3, \ldots$ developing onward with $x^\mu = (ct, \vec{x})$ at position $\vec{x}$ with time t using

$$\begin{aligned}
&{}_H\left\langle \widetilde{\Psi}(\alpha_i, t_i; t) \right| {}^H\widehat{F}^{(H)}(t, t_i) \left| \widetilde{\Psi}(\alpha_i, t_i; t) \right\rangle_H \\
&= \sum_{N=0}^{\infty} \sum_{M=0}^{\infty} \int d\widetilde{\omega}_1(t_i) \ldots d\widetilde{\omega}_N(t_i) \int d\widetilde{\omega}_1{}'(t_i) \cdots d\widetilde{\omega}_M{}'(t_i) \\
&\quad \times {}_H\langle t_i; \widetilde{\omega}_1(t_i), \ldots, \widetilde{\omega}_N(t_i), t | {}^H\widehat{F}^{(H)}(t, t_i) | t_i; \widetilde{\omega}_1'(t_i), \ldots, \widetilde{\omega}_M'(t_i), t \rangle_H \\
&\quad \times \widetilde{\Phi}_N{}^*(\alpha_i, t_i; \widetilde{\omega}_1(t_i), \ldots, \widetilde{\omega}_N(t_i), t) \widetilde{\Phi}_M(\alpha_i, t_i; \widetilde{\omega}_1'(t_i), \ldots, \widetilde{\omega}_M'(t_i), t) \\
&= \sum_{N=0}^{\infty} \sum_{M=0}^{\infty} \int d\widetilde{\omega}_1(t_i) \cdots d\widetilde{\omega}_N(t_i) \int d\widetilde{\omega}_1'(t_i) \cdots d\widetilde{\omega}_M'(t_i) \\
&\quad \times {}_S\langle t_i; \widetilde{\omega}_1(t_i), \ldots, \widetilde{\omega}_N(t_i), t_i | {}^S\widehat{F}(t) | t_i; \widetilde{\omega}_1'(t_i), \ldots, \widetilde{\omega}_M'(t_i), t_i \rangle_S \\
&\quad \times \widetilde{\Phi}_N{}^*(\alpha_i, t_i; \widetilde{\omega}_1(t_i), \ldots, \widetilde{\omega}_N(t_i), t) \widetilde{\Phi}_M(\alpha_i, t_i; \widetilde{\omega}_1'(t_i), \ldots, \widetilde{\omega}_M'(t_i), t)
\end{aligned} \tag{4.184}$$

and

$$\begin{aligned}
&{}_H\left\langle \widetilde{\Psi}(\alpha_i, t_i; t) | \widetilde{\Psi}(\alpha_i, t_i; t) \right\rangle^H H \\
&= \sum_{N=0}^{\infty} \int d\widetilde{\omega}_1(t_i) \ldots d\widetilde{\omega}_N(t_i) \left| \widetilde{\Phi}_N(\alpha_i, t_i; \widetilde{\omega}_1(t_i), \ldots, \widetilde{\omega}_N(t_i), t) \right|^2 \\
&= \sum_{N=0}^{\infty} \int d\widetilde{\omega}_1(t_i) \ldots d\widetilde{\omega}_N(t_i) \left| \widetilde{\Phi}_N(\alpha_i, t_i; \widetilde{\omega}_1(t_i), \ldots, \widetilde{\omega}_N(t_i), t_i) \right|^2
\end{aligned} \tag{4.185}$$

The time evolution of the renormalized field operator is then (see Fig. 4.16)

$$\begin{aligned}
i\hbar \frac{\partial}{\partial t} \left\langle \widetilde{\widehat{F}}(t) \right\rangle_{\alpha_i, t_i} &= \frac{{}_H\left\langle \widetilde{\Psi}(\alpha_i, t_i; t) \left| i\hbar \frac{\partial}{\partial t} \widehat{F}^{(H)}(t, t_i) \right| \widetilde{\Psi}(\alpha_i, t_i; t) \right\rangle_H}{{}_H\left\langle \widetilde{\Psi}(\alpha_i, t_i; t) | \widetilde{\Psi}(\alpha_i, t_i; t) \right\rangle_H} \\
&= \frac{{}_H\left\langle \widetilde{\Psi}(\alpha_i, t_i; t) \left| \left[\widehat{F}^{(H)}(t, t_i), \widehat{H}_{\mathrm{QED}}{}^{(H)}(t, t_i) \right] \right| \widetilde{\Psi}(\alpha_i, t_i; t) \right\rangle_H}{{}_H\left\langle \widetilde{\Psi}(\alpha_i, t_i; t) | \widetilde{\Psi}(\alpha_i, t_i; t) \right\rangle_H} \\
&\quad + \frac{{}_H\left\langle \widetilde{\Psi}(\alpha_i, t_i; t) \left| \widehat{U}^\dagger(t, t_i) \left(i\hbar \frac{\partial}{\partial t} \widehat{F}(t) \right) \widehat{U}(t, t_i) \right| \widetilde{\Psi}(\alpha_i, t_i; t) \right\rangle_H}{{}_H\left\langle \widetilde{\Psi}(\alpha_i, t_i; t) | \widetilde{\Psi}(\alpha_i, t_i; t) \right\rangle_H}
\end{aligned} \tag{4.186}$$

Fig. 4.16 Time evolution of the renormalized field operator

Field operator

$$\left\langle \tilde{\hat{F}}(t)\right\rangle_{\alpha_i,t_i} = \frac{{}_H\left\langle\tilde{\Psi}(\alpha_i,t_i;t)\right|\hat{F}^{(H)}(t,t_i)\left|\tilde{\Psi}(\alpha_i,t_i;t)\right\rangle_H}{{}_H\left\langle\tilde{\Psi}(\alpha_i,t_i;t)|\tilde{\Psi}(\alpha_i,t_i;t)\right\rangle_H}$$

$$i\hbar\frac{\partial}{\partial t}\left\langle \tilde{\hat{F}}(t)\right\rangle_{\alpha_i,t_i} = \frac{{}_H\left\langle\tilde{\Psi}(\alpha_i,t_i;t)\right|i\hbar\frac{\partial}{\partial t}\hat{F}^{(H)}(t,t_i)\left|\tilde{\Psi}(\alpha_i,t_i;t)\right\rangle_H}{{}_H\left\langle\tilde{\Psi}(\alpha_i,t_i;t)|\tilde{\Psi}(\alpha_i,t_i;t)\right\rangle_H}$$

$$= \frac{{}_H\left\langle\tilde{\Psi}(\alpha_i,t_i;t)\right|\left[\hat{F}^{(H)}(t,t_i),\hat{H}_{\mathrm{QED}}{}^{(H)}(t,t_i)\right]\left|\tilde{\Psi}(\alpha_i,t_i;t)\right\rangle_H}{{}_H\left\langle\tilde{\Psi}(\alpha_i,t_i;t)|\tilde{\Psi}(\alpha_i,t_i;t)\right\rangle_H}$$

$$+\frac{{}_H\left\langle\tilde{\Psi}(\alpha_i,t_i;t)\right|\hat{U}^{\dagger}(t,t_i)\left(i\hbar\frac{\partial}{\partial t}\hat{F}(t)\right)\hat{U}(t,t_i)\left|\tilde{\Psi}(\alpha_i,t_i;t)\right\rangle_H}{{}_H\left\langle\tilde{\Psi}(\alpha_i,t_i;t)|\tilde{\Psi}(\alpha_i,t_i;t)\right\rangle_H}$$

This concludes the way for solving the dual Cauchy problem in QED using the renormalized fields of alpha-oscillators. The renormalization has been performed over and above the alpha-resonance and thermalization. As compared with the conventional Gell–Mann–Low relationship using covariant perturbation approach (Weinberg 1995), this present approach paves the way for realizing non-perturbationally space-time resolved simulation of the time-dependent $\widehat{H}_{\mathrm{QED}}(t)$.

The interference pattern of the trajectory of $\left\langle\widetilde{\widehat{F}}(t)\right\rangle_{\alpha_i,t_i}$ *if any in the double-slit phenomenon is guaranteed primordially by that of the field operator. In other words, the interference pattern of the trajectory of* $\left\langle\widetilde{\widehat{F}}(t)\right\rangle_{\alpha_i,t_i}$ *if any has already been given prior to the assignment of the specific initial wave function* $\widetilde{\Phi}_N(\alpha_i,t_i;\widetilde{\omega}_1(t_i),\ldots,\widetilde{\omega}_N(t_i),t_i)$. *As a corollary, for different time* $t_i \neq t_j$, *even if the same initial wave function is given,* $\widetilde{\Phi}_N(\alpha_i,t_i;\widetilde{\omega}_1(t_i),\ldots,\widetilde{\omega}_N(t_i),t_i) = \widetilde{\Phi}_N(\alpha_j,t_j;\widetilde{\omega}_1(t_j),\ldots,\widetilde{\omega}_N(t_j),t_j)$, *the resultant difference* $\left\langle\widetilde{\widehat{F}}(t_{P_i})\right\rangle_{\alpha_i,t_i} \neq \left\langle\widetilde{\widehat{F}}(t_{P_j})\right\rangle_{\alpha_j,t_j}$ *is guaranteed primordially; moreover, even if a pair of events are entangled within an Einstein–Podolsky–Rosen measurement, the discrimination is guaranteed primordially*. Quantum mechanics 100 years of mystery on the measurement problem of the Minkowski space-time coordinate is solved.

It should be noted that if we were at the center of the spherical symmetric universe, particle passing through the double slit should always be connecting the image at the center of the screen. In other words, if the particles passing through the double slit are not tied to the image at the center of the screen, we should not be at the center of the spherical symmetric universe. Consequently, we observe that the double-slit phenomenon (see Fig. 1.3) guarantees that we human beings are not situated at the center of the spherical symmetric universe (see Fig. 4.17).

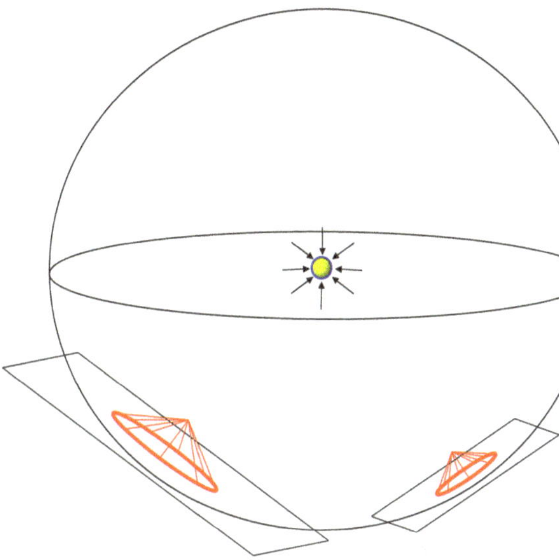

Fig. 4.17 If we were at the center of spherical symmetric universe, every particle should reach at the center, and therefore we found no single stochastic random spot away from the center even for the double-slit phenomenon

In other words, we observe that the double-slit phenomenon (see Fig. 1.3) concludes that we human beings are most probably living in the inhomogeneous universe with the inhomogeneous initial condition at $t = t_0$.

Every Boson in the standard model can be constructed by using the alpha-oscillators with the b-boson and b^c-anti-boson algebras as functions of ν-frequency, $\vec{p}$-momentum, and σ-spin; with obvious notation, read

$$\begin{aligned}\left[\hat{b}(\nu,\vec{p},\sigma),\hat{b}(\nu',\vec{q},\sigma')\right] &= \left[\hat{b}^c(\nu,\vec{p},\sigma),\hat{b}^c(\nu',\vec{q},\sigma')\right]\\ &= \left[\hat{b}^\dagger(\nu,\vec{p},\sigma),\hat{b}^\dagger(\nu',\vec{q},\sigma')\right]\\ &= \left[\hat{b}^{c\dagger}(\nu,\vec{p},\sigma),\hat{b}^{c\dagger}(\nu',\vec{q},\sigma')\right] = 0\end{aligned} \tag{4.187}$$

$$\begin{aligned}\left[\hat{b}(\nu,\vec{p},\sigma),\hat{b}^\dagger(\nu',\vec{q},\sigma')\right] &= \left[\hat{b}^c(\nu,\vec{p},\sigma),\hat{b}^{c\dagger}(\nu',\vec{q},\sigma')\right]\\ &= \delta_{\sigma\sigma'}\delta^3(\vec{p}-\vec{q})\delta\big(\nu-\nu(|\vec{p}|)_b\big)\delta\big(\nu'-\nu(|\vec{q}|)_b\big)\end{aligned} \tag{4.188}$$

In due course, the alpha-resonance condition, the dispersion relationships, and the coarse graining should also be met.

Likewise, every fermion can be constructed by using the alpha-oscillators with the f-fermion and f^c-anti-fermion algebras as functions of ν-frequency, $\vec{p}$-momentum, and σ-spin; with obvious notation, read

$$\begin{aligned}\left\{\hat{f}(\nu,\vec{p},\sigma),\hat{f}(\nu',\vec{q},\sigma')\right\} &= \left\{\hat{f}^c(\nu,\vec{p},\sigma),\hat{f}^c(\nu',\vec{q},\sigma')\right\}\\ &= \left\{\hat{f}^\dagger(\nu,\vec{p},\sigma),\hat{f}^\dagger(\nu',\vec{q},\sigma')\right\}\\ &= \left\{\hat{f}^{c\dagger}(\nu,\vec{p},\sigma),\hat{f}^{c\dagger}(\nu',\vec{q},\sigma')\right\} = 0\end{aligned} \tag{4.189}$$

$$\left\{\widehat{f}(\nu,\vec{p},\sigma),\widehat{f}^{\dagger}(\nu',\vec{q},\sigma')\right\} = \left\{\widehat{f}^{c}(\nu,\vec{p},\sigma),\widehat{f}^{c\dagger}(\nu',\vec{q},\sigma')\right\}$$
$$= \delta_{\sigma\sigma'}\delta^{3}(\vec{p}-\vec{q})\delta\left(\nu-\nu(|\vec{p}|)_{f}\right)\delta\left(\nu'-\nu(|\vec{q}|)_{f}\right) \tag{4.190}$$

In due course, the alpha-resonance condition, the dispersion relationships, and the coarse graining should also be met.

Super alpha-oscillator algebra that granted SUSY as the local symmetry of the alpha-oscillator theory gives the graviton of SUGRA.

Finally, it should be noted that in general, the alpha-resonance condition and/or the dispersion relationships and/or the coarse graining may not be met; with obvious notation, read

$$cp^{0}\left(\nu(|\vec{p}|)_{b,f},|\vec{p}|\right) \neq h\nu(|\vec{p}|)_{b,f},\ c\sqrt{\left(m_{b,f}c\right)^{2}+|\vec{p}|^{2}} \tag{4.191}$$

$$\int_{0}^{\infty}d\nu\int_{0}^{\infty}d\nu'\left[\widehat{\alpha}(\nu),\widehat{\beta}(\nu')\right]_{\pm} \neq \left[\widehat{\alpha}_{\text{particle}},\widehat{\beta}_{\text{particle}}\right]_{\pm} \tag{4.192}$$

In other words, in the generic situation, we have dense set of ket vectors out of the coarse graining, since normal modes are immersed in alpha-oscillators (see Fig. 4.18).

$$\{\text{alpha-oscillators}\} \supset \{\text{normal modes}\} \tag{4.193}$$

We may call the very basic energy as the alpha-oscillator energy as compared with the particle energy. *The alpha-oscillator energy may contribute to the total energy but not as the known form of particles.* In this sense, the alpha-oscillator theory may give a natural candidate of dark energy, which is said to be abundant in our universe not as the known form of particles and to account for the cause of the accelerated expansion of our universe (see Fig. 4.19).

Moreover, the time evolution of ket vector with wave function of the thermalized alpha-oscillators, Eq. (4.101), may be associated with the time evolution of ket vector with wave function of particles, Eq. (4.149). This situation may be called an alpha-weighted state. The time evolution of ket vector with wave function of the alpha-weighted state is given for an event α_i starting at t_i with $t_0 < t_i$; $i = 1, 2, 3, \ldots$, with obvious notation as

$$\begin{aligned}
&\left|\widetilde{\Psi}^{\text{Alpha}}(\alpha_i, t_i; t)\right\rangle_{H\text{ or }S} \\
&= \sum_{N=0}^{\infty}\int\left\{d\widetilde{\omega}_1(t_i)\cdots d\widetilde{\omega}_{N_{\widetilde{\omega}}}(t_i)\right\}\otimes\{d\omega_1\cdots d\omega_{N_\omega}\}_{\omega\neq\widetilde{\omega}} \\
&\quad\times\left|t_i;\{\widetilde{\omega}_{11}(t_i),\ldots,\widetilde{\omega}_{N_{\widetilde{\omega}}}(t_i)\}\otimes\{\omega_1,\ldots,\omega_{N_\omega}\}_{\omega\neq\widetilde{\omega}},t_i\right\rangle_{H\text{ or }S} \\
&\quad\times\widetilde{\Phi}_{N}{}^{\text{Alpha}}\left(\alpha_i,t_i;\left\{\widetilde{\omega}_1(t_i),\ldots,\widetilde{\omega}_{N_{\widetilde{\omega}}}(t_i)\right\}\otimes\{\omega_1,\ldots,\omega_{N_\omega}\}_{\omega\neq\widetilde{\omega}},t\right)
\end{aligned} \tag{4.194}$$

$$N = N_{\omega}\oplus N_{\widetilde{\omega}},\quad \omega\neq\widetilde{\omega} \tag{4.195}$$

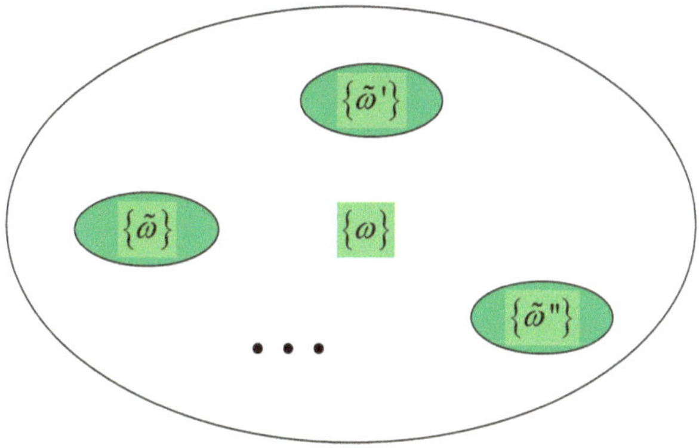

Fig. 4.18 Normal modes $\{\tilde{\omega}\}, \{\tilde{\omega}'\}, \{\tilde{\omega}''\}, \ldots$ are immersed in alpha-oscillators $\{\omega\}$

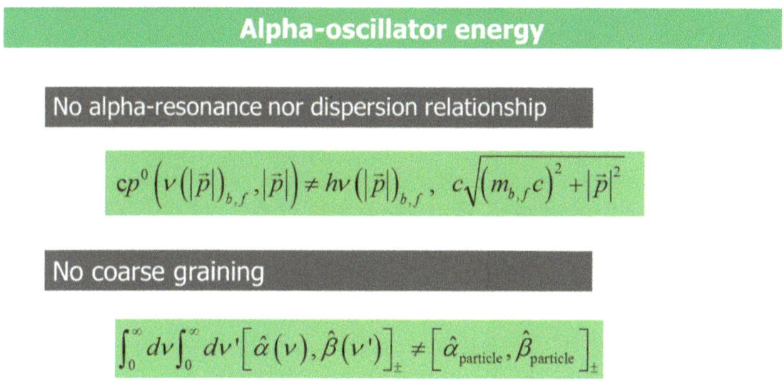

Fig. 4.19 Alpha-oscillator energy may contribute to the total energy but not as the known form of particles

In this alpha-weighted state, we have the time evolution of wave functions with the obvious notation

$$
\begin{aligned}
& i\hbar\frac{\partial}{\partial t}\widetilde{\Phi}_N{}^{\text{Alpha}}\left(\alpha_i, t_i; \{\widetilde{\omega}_1(t_i), \ldots, \widetilde{\omega}_{N_{\tilde{\omega}}}(t_i)\} \otimes \{\omega_1, \ldots, \omega_{N_\omega}\}_{\omega \neq \tilde{\omega}}, t\right) \\
& = \sum_{M=0}^{\infty}\int \left\{d\widetilde{\omega}'_1(t_i) \ldots d\widetilde{\omega}'_{N_{\tilde{\omega}'}}(t_i)\right\} \otimes \left\{d\omega'_1 \cdots d\omega'_{N_{\omega'}}\right\}_{\omega' \neq \tilde{\omega}'} \\
& \quad \times H_{NM}\Big(t_i; \{\widetilde{\omega}_1(t_i), \ldots, \widetilde{\omega}_{N_{\tilde{\omega}}}(t_i)\} \otimes \{\omega_1, \ldots, \omega_{N_\omega}\}_{\omega \neq \tilde{\omega}}, \left\{\widetilde{\omega}'_1(t_i), \ldots, \widetilde{\omega}'_{M_{\tilde{\omega}'}}(t_i)\right\} \\
& \quad \otimes \left\{\omega'_1, \ldots, \omega'_{M_{\omega'}}\right\}_{\omega' \neq \tilde{\omega}'}, t\Big) \\
& \quad \times \widetilde{\Phi}_M{}^{\text{Alpha}}\left(\alpha_i, t_i; \left\{\widetilde{\omega}'_1(t_i), \ldots, \widetilde{\omega}'_{M_{\tilde{\omega}'}}(t_i)\right\} \otimes \left\{\omega'_1, \ldots, \omega'_{M_{\omega'}}\right\}_{\omega' \neq \tilde{\omega}'}, t\right)
\end{aligned}
\tag{4.196}
$$

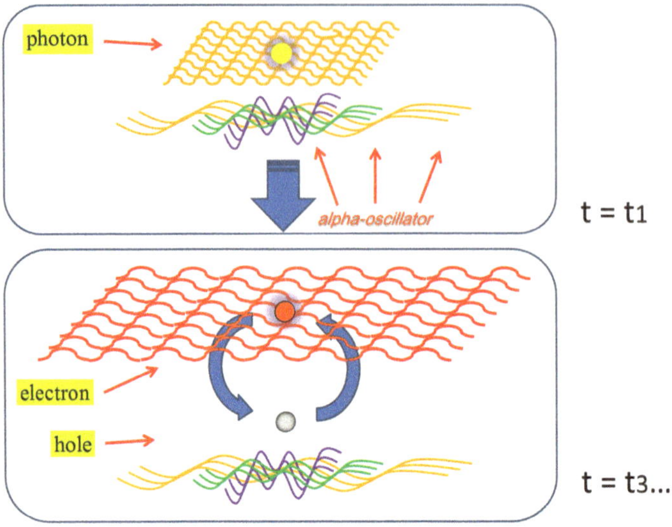

Fig. 4.20 The alpha-weighted state allows the thermalized alpha-oscillators be realized in the time evolution of ket vector with wave function, when sub-particles be interchangeable with particles in (i) exciton pair formation, (ii) photoelectric effect, and (iii) electron–positron pair production (the hole should read positron in this last case)

For example, exchange of sub-particles with particles may be described by the alpha-weighted state when the particle number changes. Most typical candidates are the photon number nonconservation cases, such as the exciton pair formation (see Fig. 4.8), photoelectric effect, and electron–positron pair production (see Fig. 4.20).

As shown in Fig. 4.20, photon disappears from the wave function of particle but can return to that of the alpha-oscillator in the form of sub-particle, *b*-photon, when electron is irradiated to change the wave function to that of the specific excited state. In the case with electron–positron pair production, the electron and positron can emerge from the wave functions of sub-particles, *f*-electron andf^c-positron, respectively. The time evolution of the field operator, no matter whether it is renormalized or not, in the alpha-weighted state is then (see Fig. 4.21)

$$\left\langle \widetilde{\widehat{F}}^{\text{Alpha}}(t) \right\rangle_{\alpha_i, t_i} = \frac{{}_H\left\langle \widetilde{\Psi}^{\text{Alpha}}(\alpha_i, t_i; t) \middle| \widehat{F}^{(H)}(t, t_i) \middle| \widetilde{\Psi}^{\text{Alpha}}(\alpha_i, t_i; t) \right\rangle_H}{{}_H\left\langle \widetilde{\Psi}^{\text{Alpha}}(\alpha_i, t_i; t) \middle| \widetilde{\Psi}^{\text{Alpha}}(\alpha_i, t_i; t) \right\rangle_H} \tag{4.197}$$

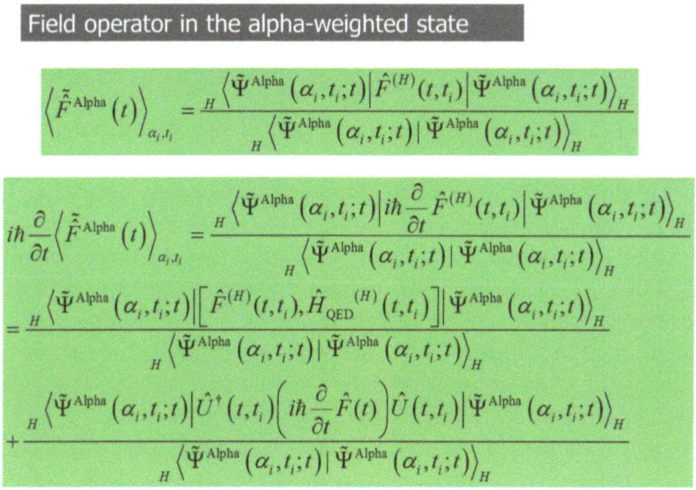

Fig. 4.21 Time evolution of field operator, no matter whether it is renormalized or not, in the alpha-weighted state

$$
\begin{aligned}
i\hbar\frac{\partial}{\partial t}\left\langle \widetilde{\widehat{F}}^{\text{Alpha}}(t)\right\rangle_{\alpha_i,t_i} &= \frac{{}_H\left\langle \widetilde{\Psi}^{\text{Alpha}}(\alpha_i,t_i;t)\middle| i\hbar\frac{\partial}{\partial t}\widehat{F}^{(H)}(t,t_i)\middle|\widetilde{\Psi}^{\text{Alpha}}(\alpha_i,t_i;t)\right\rangle_H}{{}_H\left\langle \widetilde{\Psi}^{\text{Alpha}}(\alpha_i,t_i;t)|\widetilde{\Psi}^{\text{Alpha}}(\alpha_i,t_i;t)\right\rangle_H} \\
&= \frac{{}_H\left\langle \widetilde{\Psi}^{\text{Alpha}}(\alpha_i,t_i;t)\middle|\left[\widehat{F}^{(H)}(t,t_i),{\widehat{H}_{\text{QED}}}^{(H)}(t,t_i)\right]\middle|\widetilde{\Psi}^{\text{Alpha}}(\alpha_i,t_i;t)\right\rangle_H}{{}_H\left\langle \widetilde{\Psi}^{\text{Alpha}}(\alpha_i,t_i;t)|\widetilde{\Psi}^{\text{Alpha}}(\alpha_i,t_i;t)\right\rangle_H} \\
&+ \frac{{}_H\left\langle \widetilde{\Psi}^{\text{Alpha}}(\alpha_i,t_i;t)\middle|\widehat{U}^{\dagger}(t,t_i)\left(i\hbar\frac{\partial}{\partial t}\widehat{F}(t)\right)\widehat{U}(t,t_i)\middle|\widetilde{\Psi}^{\text{Alpha}}(\alpha_i,t_i;t)\right\rangle_H}{{}_H\left\langle \widetilde{\Psi}^{\text{Alpha}}(\alpha_i,t_i;t)|\widetilde{\Psi}^{\text{Alpha}}(\alpha_i,t_i;t)\right\rangle_H}
\end{aligned}
\tag{4.198}
$$

More specifically in Fig. 4.20, if the field operator $\widehat{F}(t)$ is that of photon in the double-slit phenomenon, then the expectation value $\left\langle \widetilde{\widehat{F}}^{\text{Alpha}}(t)\right\rangle_{\alpha_1,t_1}$ has finite value at $t=t_1$ around the first slit, but zero value at $t=t_3$ around the screen later. In the case of electron–positron pair production in Fig. 4.20, if the field operator $\widehat{F}(t)$ is that of photon, then the expectation value $\left\langle \widetilde{\widehat{F}}^{\text{Alpha}}(t)\right\rangle_{\alpha_1,t_1}$ has finite value at $t=t_1$ around the point of the electron–positron pair production, but zero value at $t=t_3$ around there later; also, if the field operator $\widehat{F}(t)$ is that of electron or positron, then the expectation value $\left\langle \widetilde{\widehat{F}}^{\text{Alpha}}(t)\right\rangle_{\alpha_1,t_1}$ has finite value at $t=t_3$ around the point of the electron–positron pair production, but zero value at $t=t_1$ around there earlier.

4.6 Summary

We have generalized the time evolution operator machinery of quantum mechanics to QED in such a way that it warrants the use of the time-dependent $\widehat{H}_{\mathrm{QED}}(t)$. Non-perturbational space-time-resolved simulation of QED has been realized in terms of the dual Cauchy problem. The alpha-resonance condition has been restricted to the dispersion relationship of particles for time-dependent renormalization of the alpha-oscillators. The commutation relationships of particles are obtained by coarse graining of the alpha-oscillators. The alpha-oscillator energy may contribute to the total energy but not as the known form of particles.

The alpha-oscillator theory with the alpha-weighted state works well for non-perturbationally space-time-resolved simulation of the time-dependent nonconservation of particle number. *The nonconservation of particle number is in the heart of quantum field theory like QED that describes nature as is.* The conventional putative time-independent EMstatic Hamiltonian $\widehat{H}_{\mathrm{EMstatic\ QED}}$ fails to account for the particle number nonconservation. In quantum mechanics also, using the conventional time-independent electrostatic Hamiltonian $\widehat{H}_{\mathrm{electrostatic}}$, every particle number is conserved, so that mysterious "collapse of wave packet" should be invoked instead.

References

R.P. Feynman, R.B. Leighton, M. Sands, *The Feynman Lectures on Physics* (Addison-Wesley, Reading, 1965) Vol. III, Chap. 1

L. Guzzo et al., Nature **451**, 541 (2008)

L.D. Landau, E.M. Lifshitz, *The Classical Theory of Fields*, 6th edn. (Nauka, Moscow, 1973), translated into English in *Course of Theoretical Physics Series, Volume* **2**, 4th edn. (Pergamon, Oxford, 1975)

N. Nakanishi, Prog. Theor. Phys. **111**, 301 (2004)

J.J. Sakurai, *Modern Quantum Mechanics* (The Benjamin/Cummings, Menlo Park, 1985)

A. Tachibana, in *Concepts and Methods in Modern Theoretical Chemistry: Electronic Structure and Reactivity (Atoms, Molecules, and Clusters)*, ed by S. K. Ghosh, P. K. Chattaraj. (Taylor & Francis/CRC, New York, 2013) Chap.12, pp. 235–251

A. Tachibana, J. Math. Chem. **53**, 1943 (2015)

A. Tachibana, J. Math. Chem. **54**, 661 (2016)

A. Tonomura, Proc. Natl. Acad. Sci. USA **102**, 14952 (2005)

S. Weinberg, *The Quantum Theory of Fields* (Cambridge University, Cambridge, 1995)

Index

A. Tachibana, *New Aspects of Quantum Electrodynamics*,
DOI 10.1007/978-981-10-3132-8

Kolibri 360 GmbH
Ferdinand-Jühlke-Straße 7
99095 Erfurt, Deutschland
produktsicherheit@kolibri360.de